LABORATORY MANUAL

PRINCIPLES OF

ELECTRONIC COMMUNICATION SYSTEMS

DAVID L. HEISERMAN

Glencoe McGraw-Hill

New York, New York   Columbus, Ohio   Mission Hills, California   Peoria, Illinois

Cover Photo Credit: Myron J. Dorf/The Stock Market, (l) Dale O'Dell/The Stock Market; (r) David Lawrence/ The Stock Market.

Laboratory Manual for
Principles of Electronic Communication Systems

ISBN: 0-02-800410-8

Glencoe software support: 1 800 437 3715

Printed in the United States of America.

2 3 4 5 6 7 8 9   045   01 00 99 98 97

Computer Requirements and General Instructions

Hardware Requirements

- IBM PC (or equivalent) 386 SX or higher; 486 is preferred.
- 4 MB RAM, minimum.
- 4 MB available on the hard drive.
- Windows 3.1 or higher
- VGA or better.
- Sound optional
- Internet access optional

Installation

The Comm Lab software must be installled from the two diskettes onto your hard drive. However, you do not need to repeat this process once you have completed the following five-step procedure:

1. Start Microsoft Windows on the computer.

2. Insert the *Comm Lab* diskette 1 into drive A (or B).

3. From the Program Manager, select the File menu and choose Run.

4. Type **A:\SETUP** (or **B:\SETUP**) and press **Enter**.

5. Follow the progress on the screen.

To start the program any time after installation, double-click the *Comm Lab 1.0* icon in the Comm Lab group of your Program Manager.

Using Windows

This manual describes how to carry out simulated laboratory projects — where to set the dials and how to set the dials, for example. We assume that you already know the basic Windows procedures such as using scroll bars, clicking and double-clicking the mouse, and performing keyboard shortcuts. Windows 3.1 users can get help with these fundamentals by clicking the Windows Tutoral icon in the Comm Lab group of the Program Manager.

Simulated Controls

Most simulated projects require you to set signal amplitudes and frequencies. The control that is used most often for this prupose is the Windows slide control. You can make very coarse adjustments by dragging the slider button to the desired position. You can make orderly stepwise adjustments by clicking the slider bar on either side of the slider button; and you can make very fine adjustments by clicking one of the arrow buttons located at the ends of the slider.

Addional hints and helps for other kinds of controls are always available from the simulated projects, themselves. Hints for using the simulated instruments are available by setting the mouse over the instrument in question and pressing the right mouse button. Detailed helps are available by clicking the upper-left corner of an instrument with the left mouse button.

Projects: Simulated and Hands-On

This lab manual contains both computer-simulated and hands-on projects. The simulated projects are performed in a computer lab setting, and the hands-on projects are to be peformed at a real workbench with real circuits and test equipment.

In the table of contents that follows on page vii, computer simulated projects are marked with a computer icon.

Hands-on projects are indicated with the figure of a hand holding a test probe to a circuit.

About the Author

David L. Heiserman is a fifteen-year veteran of the class-room, conducting classes in electronics technology and technical mathematics at DeVry, Franklin University (Columbus, Ohio), and Columbus State Technical College.

He earned his degree at Ohio State University in applied mathematics and psychology. His knowledge of electronics stems from his years as an aviation radar control systems specialist in the Navy. He holds two U.S. patents for robotic applications of piezoelectric polymers.

He has authored over thirty books on technical and scientific subjects, including three on digital electronics and microprocessor instruction sets.

Preface

The domain of modern electronic communications is expanding rapidly — into realms of higher frequencies and narrower bandwidths, more and better consumer products, and entirely new communications systems. And just as the technology is becoming more sophisticated and wider ranging, so must the methods for preparing the engineers and technicians who make it all work. This laboratory manual, and its accompanying simulation software and Internet support, represent a significant step in that direction.

Sixty years ago, the most sophisticated and expensive radio-frequency equipment could operate only in the hundreds of megahertz range. Today, however, there are consumer wireless devices now operating in the thousands of megahertz, or gigahertz, range. Just ten years ago, the telephone was a relatively simple instrument that carried low-frequency, narrow-channel voice signals. Now the same device and its supporting system is expected to handle voice, fax, and computer telecommunications.

Communications technology was once fully occupied with the transmission and reception of voice and pictures, but is now becoming the technology of telecommunications. Millions of personal computers in homes, schools, and businesses are interconnected on local networks and over the entire world via the Internet. Communications between computers is no longer an option; rather, it is a vital part of the development of computer technology.

So it is clear that communications technology is indeed growing and changing very rapidly; and it is equally evident that the methods and resources for teaching communications technology must keep pace with these changes. This laboratory system (including the hardcopy manual and software projects) answers this demand by offering a far wider range and number of projects than possible with the traditional approach. The sequence of projects follows the textbook, chapter-by-chapter, supporting the theory and expanding the students' experiences to include computer simulations as well as direct, hands-on projects with communications circuits, devices, and test equipment.

Most of the projects included in this book require the student to interact with simulated equipment, circuits, and devices located on the computer screen. Many of these are *prep projects*—projects that prepare the student for more effective use of limited lab time. Every prep project precedes a *hands-on* project. The hands-on projects are to be conducted in the laboratory, using conventional instruments and procedures. (See Appendix A for a complete listing of components, supplies, and equipment required for the hands-on projects.)

In conventional, hands-on communications labs, most time is spent connecting the circuits and adjusting the test instruments. Once the circuit is operating and the test instruments are adjusted properly, little time remains for thoughtful and thorough experimentation. The prep projects in this laboratory system eliminate the need for constructing the circuit and struggling with equipment adjustments. The projects allow the student to concentrate upon the tasks of gathering data and, more importantly, observing the way a circuit responds to changing input conditions. But having gained this valuable experience with the operating principles of a circuit, the student can then move to the corresponding hands-on lab where he or she can afford to give full attention to constructing actual circuits and adjusting real instruments.

Few school laboratories can offer ready access to the full range of instruments and circuits that represent the new domains of communications technology. Hands-on projects in such instances are entirely impractical. But the laboratory system you are using here can simulate lab experiences with sophisticated, expensive, and dangerous equipment. This is the purpose of the *Extended* projects.

The simulated projects all use dynamic, interactive, graphical formats; the student can adjust input parameters to a test circuit and view the results in real time. These projects do not rely on the cumbersome and unrealistic "keyboard ergonomics" of previous generations of computer simulations. Voltage and frequency sources, for example, are adjusted by using the mouse to grasp and slide simulated controls located on an image of the instrument. The output values respond instantly, usually in a digital format, on simulated displays. Outputs from the circuits also are displayed in real time on simulated instruments. These output devices often use a digital format, but analog displays are used frequently enough to help the student feel comfortable with them (and note their advantages under certain types of test conditions).

The projects in this manual follow closely the material presented in Louis E. Frenzel's *Principles of Electronic Communication Systems*. Each lab assignment, for example, references relevant sections in the textbook. Also, the nomenclature, abbreviations and mathematical notations in this manual are consistent with those presented in the textbook. There are far more hours of laboratory work presented here than is normally available in the electronics and computer labs. The instructor can draw upon a selection of hands-on and simulated lab projects deemed most important for supporting the pace and content of the lecture. Also, most simulated labs are suitable for the instructor to demonstration in the classroom.

Software support is available by telephone at

1-800-437-3715

and on the World Wide Web at

http://www.glencoe.com/

The author maintains a World Wide Web site devoted to hints and further suggestions at

http://www.infinet.com/~sweethvn

The student of communications electronics traditionally sits on a lab stool at a bench on which there are a few pieces of lab equipment and a handful of transistors, coils, and other small components. With the laboratory system used in this book and the accompanying software, the environment is expanded to include the simulation effects of a computer and the information resources of the Internet. Welcome to education in the twenty-first century!

David L. Heiserman

Contents

Project 1

Passive Filters I

A Prep Project

This project is a computer simulation of laboratory tests for determining the frequency response of passive low-pass and high-pass *RC* filters. For both types of circuits you will:

- Calculate the cut-off frequency.
- Gather data for the response curve.
- Plot the response curve on a semilog graph.
- Determine the cut-off frequency from the curve.
- Confirm the actual cut-off frequency by direct measurement.

Preparation

Read Frenzel, *Principles of Electronic Communication Systems*, Section 2-3.

Setup Procedure

1. Select Prep Projects from the Project menu.

2. Select Project 1 Passive Filters I from the list of Prep Projects.

Lab Procedure

For both parts of this project you will be using the simulated function generator and ac voltmeter. The function generator supplies the required input waveform, and the ac voltmeter provides a convenient means for monitoring the output voltage level.

The schematic diagrams accurately represent the circuit under test. The block diagrams indicate how the equipment and circuit are interconnected.

Part 1   Low-Pass *RC* Filter

Check the project bar across the top of the screen to confirm you are working with Part 1 of this project.

1. Calculate the cut-off frequency for the low-pass *RC* filter circuit shown on the screen as the schematic diagram for Part 1. Record the values of R and C, and your calculated value of f_{co} on the Results Sheet for this project.

2. Adjust the amplitude of the function generator for 10.0 V_{p-p}. Given this 10-volt input, calculate the amount of voltage that should appear at the output of the circuit when the frequency is adjusted to f_{co}. Record the calculated value on the Results Sheet.

3. Adjust the frequency of the function generator to each of the values shown in Table 1-1 on the Results Sheet. In each case, record the value of v_o as shown on the ac voltmeter.

4. Set the frequency of the function generator to the value of f_{co} that you calculated in Step 1. While observing the output on the ac voltmeter, adjust the frequency slightly above and below the calculated value of f_{co}. Note the frequency value that causes the output voltage to equal the cut-off voltage that you calculated in Step 3. Record this frequency as measured f_{co} on the Results Sheet.

5. Calculate the dB response of the circuit for each of the measurements in Table 1-1, then plot the data on the semilog graph shown as Figure 1-1 on the Results Sheet. Also include a point for your measured value of f_{co}.

Click the browse button to go to Part 2 of this project.

Part 2   High-Pass *RC* Filter

Check the screen's project bar to confirm you are working with Part 2 of this project.

1. Calculate the cut-off frequency for the high-pass *RC* filter circuit shown on the screen as the schematic diagram for Part 2. Record the values of *R* and *C*, and your calculated value of f_{co} on the Results Sheet for Project 1.

2. Adjust the amplitude of the function generator for 10.0 $V_{p\text{-}p}$. Given this 10-volt input, calculate the amount of voltage that should appear at the output of the circuit when the frequency is adjusted to f_{co}. Record the calculated v_o on the Results Sheet.

3. Adjust the frequency of the function generator to each of the values shown in Table 1-2 on the Results Sheet. In each case, record the value of v_o as shown on the ac voltmeter.

4. Set the frequency of the function generator to the value of f_{co} that you calculated in Step 1. While observing the output on the ac voltmeter, adjust the frequency slightly above and below the calculated value of f_{co}. Note the frequency value that causes the output voltage to equal the cut-off voltage that you calculated in Step 3. Record this frequency as measured f_{co} on the Results Sheet.

5. Calculate the dB response of the circuit for each of the measurements in Table 1-2. Plot this output data on the semilog graph shown in Figure 1-2 on the Results Sheet. Also include a point for your measured value of f_{co}.

Click the exit button when you are ready to leave this project.

Results Sheet

Project 1

Part 1   Low-Pass *RC* Filter

Step 1

  $R =$ _____ $C =$ _____

  Calculated $f_{co} =$ _____

Step 2

  Calculated $v_o =$ _____

Step 4

  Measured $f_{co} =$ _____

f kHz	v_o V_{p-p}	dB Response*
85.0		
95.0		
150.0		
250.0		
350.0		
450.0		
550.0		
650.0		
750.0		
850.0		

$$* \; dB \; response \; = 20 \log \frac{v_o}{v_{in}}$$

Table 1-1

Questions

1. Describe how your curve in Figure 1-1 indicates that the circuit is operating as a low-pass filter.

2. What is the approximate roll-off rate of your curve in terms of dB/decade?

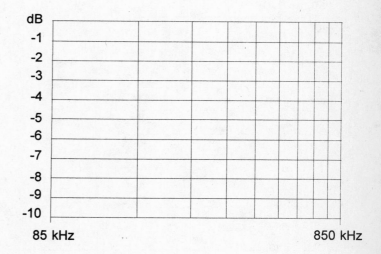

Figure 1-1

Part 2   High-Pass *RC* Filter

Step 1

R = _____   C = _____

Calculated f_{co} = _____

Step 2

Calculated v_o = _____

Step 4

Measured f_{co} = _____

f kHz	v_o V_{p-p}	dB Response
85.0		
95.0		
150.0		
250.0		
350.0		
450.0		
550.0		
650.0		
750.0		
850.0		

Table 1-2

Questions

1. Describe how your curve in Figure 1-2 indicates that the circuit is operating as a high-pass filter.

2. What is the approximate roll-off rate of your curve in terms of dB/decade?

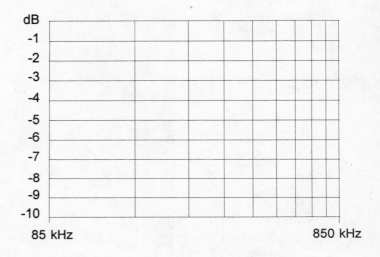

Figure 1-2

Critical Thinking for Project 1

1. Describe how increasing the value of *C* in an *RC* low-pass filter affects the circuit's cut-off frequency. Compare this with increasing the value of *C* in an *RC* high-pass filter.

2. Describe the effect that a shorted capacitor has upon the response curve of an *RC* low-pass filter. Compare this with the effect of an open capacitor in an *RC* high-pass filter.

3. Derive a formula for calculating the maximum and minimum cut-off values for an *RC* low-pass filter when the value of *R* can be varied from R_{min} to R_{max}.

Project 2

Passive Filters I

A Hands-On Project

In this project you will study passive low- and high-pass *RC* filter circuits. For each type of circuit you will:

- Construct the circuit.
- Calculate the cut-off frequency.
- Gather data for the response curve.
- Plot the response curve on a semilog graph.
- Determine the cut-off frequency from the curve.
- Confirm the actual cut-off frequency by direct measurement.

Preparation

Read Frenzel, *Principles of Electronic Communication Systems*, Section 2-3.

Complete the work for Prep Project 1.

Components and Supplies

1   Resistor, 4.7 kΩ
1   Capacitor, 10 nF

Equipment

1   Function generator
1   Dual-trace oscilloscope
1   Frequency counter (optional)

The frequency counter is optional because it is possible to use the oscilloscope to determine the circuit's operating frequency.

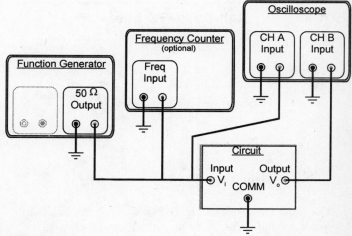

Figure 2-1

Lab Procedure

Part 1   Low-Pass Filter

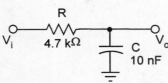

Figure 2-2

1. Construct the circuit of Figure 2-2. Connect the function generator and oscilloscope as shown in Figure 2-1.

2. For each of the entries in Table 2-1 on the Results Sheet, set the function generator for the given frequency and the output voltage to 10 $V_{p\text{-}p}$. Measure and record the peak-to-peak value of v_o.

3. Determine the maximum value of v_o from the data in Table 2-1. Use this value as the 0 dB level for converting all other values for v_o in Table 2-1 to their corresponding dB level. Plot the results on the semilog graph of Figure 2-4.

4. Calculate the cut-off frequency, f_{co}, according to the values for R and C in Figure 2-2. Estimate the cut-off frequency from the data in your graph. Show your values on the Results Sheet.

5. Determine the -3 dB voltage level for this circuit by multiplying the maximum output voltage by 0.707. Record this value on the Results Sheet. Make sure v_i to the circuit is still 10 $V_{p\text{-}p}$, and adjust the frequency to the actual cut-off frequency (to the point where v_o is the voltage you calculated previously in this step). Record the frequency as the measured cut-off frequency.

Part 2   High-Pass *RC* Filter

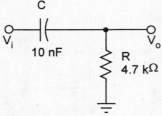

Figure 2-3

1. Construct the circuit of Figure 2-3. Connect the function generator and oscilloscope as in the previous parts of this project (Figure 2-1).

2. For each of the entries in Table 2-2, set the function generator for the given frequency, double-check the value of v_i to make sure it is 10 $V_{p\text{-}p}$, and record the peak-to-peak value of v_o.

3. Determine the maximum value of v_o from the data in Table 2-2. Use this value as the 0 dB level for converting all other values for v_o in Table 2-2 to their corresponding dB levels. Plot the results on the semilog graph of Figure 2-5.

4. Calculate the cut-off frequency, f_{co}, according to the values of R and C in this circuit. Estimate the cut-off frequency from the graph. Record these values on the Results Sheet.

5. Determine the -3 dB voltage level for this circuit as you did for Step 5 of Part 1. Record this value on the Results Sheet. Make sure the voltage applied at v_i is still 10 $V_{p\text{-}p}$, and adjust the frequency to the actual cut-off frequency (to the point where v_o is the voltage you calculated previously in this step). Record the frequency as the measured cut-off frequency.

Results Sheet

Project 2

Part 1   Low-Pass *RC* Filter

f kHz	v_o V p-p	$v_o / v_{i(max)}$ V p-p	dB Response
1			
2			
3			
4			
5			
6			
7			
8			
9			

Table 2-1

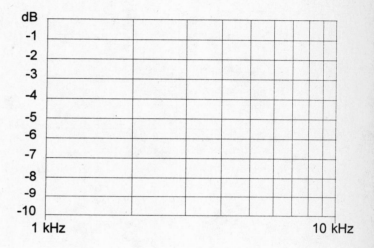

Figure 2-4

Step 4

   Calculated f_{co} = _____

   Estimated f_{co} = _____

Step 5

   v_o at −3 dB = _____

   Measured f_{co} = _____

Questions

1. What is the approximate roll-off rate (in dB/decade) of this circuit?

2. What would happen to the value of f_{co} if the value of C were increased?

3. What would happen to the actual value of f_{co} if the value of v_i were decreased?

7

Part 2   High-Pass *RC* Filter

f kHz	v_o $V_{p\text{-}p}$	$v_o / v_{i(max)}$	dB Response
1			
2			
3			
4			
5			
6			
7			
8			
9			

Table 2-2

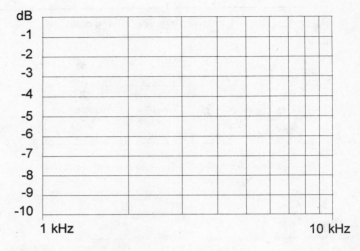

Figure 2-5

Step 4

    Calculated f_{co} = _____

    Estimated f_{co} = _____

Step 5   /

    v_o at -3 dB = _____

    Measured f_{co} = _____

Questions

1. What is the approximate roll-off rate of this circuit (expressed in dB/decade)?

2. What would happen to the value of f_{co} if the value of C were increased?

3. What would happen to the value of f_{co} if the actual value of v_i were increased?

Critical Thinking for Project 2

1. Capacitive coupling is often used between stages of audio amplifiers. Explain why this use causes difficulties with low-frequency audio response.

2. Sketch the schematic diagram of a low-pass *LR* filter circuit.

3. Sketch the schematic diagram of a high-pass *LR* filter circuit.

4. Describe a practical reason why *RC* combinations are more often used than *LR* combinations in high- and low-pass filter circuits.

Project **3**

Passive Filters II

A Prep Project

This project is a computer simulation of laboratory tests for determining the frequency response of passive bandpass and band-reject filters. For both types of circuits you will:

- Calculate the cut-off frequency.
- Gather data for the response curve.
- Plot the response curve on a semilog graph.
- Determine the center, upper, and lower cut-off frequencies from the curve.
- Confirm the actual frequencies by direct measurement.

Preparation

Read Frenzel, *Principles of Electronic Communication Systems*, Section 2-3.

Setup Procedure

1. Select Prep Projects from the Project menu.

2. Select Project 3 Passive Filters II from the list of Prep Projects.

Lab Procedure

For both parts of this project you will be using the simulated function generator and ac voltmeter. The function generator supplies the required input waveform. The ac voltmeter provides a convenient means for monitoring the output voltage level.

The schematic diagrams exactly represent the circuit under test. The block diagrams indicate how the equipment and circuit are interconnected.

Part 1   Series *LC* Bandpass Filter

Check the project bar across the top of the screen to confirm you are working with Part 1 of this project.

1. Note on the screen the schematic diagram for Part 1. Record the values of L and C on the Results Sheet for Project 3. Calculate and record the following on the Results Sheet: the circuit's center frequency (f_c), the lower -3 dB cut-off frequency (f_1), and the upper cut-off frequency (f_2).

2. Adjust the amplitude of the function generator for 10.0 $V_{p\text{-}p}$. Adjust the frequency of the function generator to each of the values shown in Table 3-1 on the Results Sheet. In each case, record the value of v_o as shown on the ac voltmeter.

3. Set the frequency of the function generator to the value of f_c that you calculated in Step 1. While observing the output on the ac voltmeter, adjust the frequency slightly above and below the calculated value of f_c. Note the frequency value that causes the **maximum** amount of output voltage. Record this frequency on the Results Sheet.

4. Set the frequency of the function generator to the value of f_1 that you calculated in Step 1. While observing the output on the ac voltmeter, adjust the frequency slightly above and below the calculated value until the output voltage equals 0.707 times the maximum output voltage. Record this measured value of f_1 on the Results Sheet. Repeat the adjustments and measurements for the upper cut-off point, f_2.

5. Calculate the dB voltage gain of the circuit for each of the measurements in Table 3-1. Plot this output data on the semilog graph shown in Figure 3-1 on the Results Sheet. Also include the points for measured values of f_c, f_1, and f_2 from Steps 3 and 4 of this procedure.

Click the browse button to go to Part 2 of this project.

Part 2　　Series *LC* Band-reject Filter

Check the screen's project bar to confirm you are working with Part 2 of this project.

1. Record on the Results Sheet the values of L and C shown on the schematic for Part 2. Calculate and record the circuit's values for f_c, f_1, and f_2.

2. Adjust the amplitude of the function generator for 10.0 V_{p-p}. Adjust the frequency of the function generator to each of the values shown in Table 3-2 on the Results Sheet. In each case, record the value of v_o as shown on the ac voltmeter.

3. Set the frequency of the function generator to the value of f_c that you calculated in Step 1. While observing the output on the ac voltmeter, adjust the frequency slightly above and below the calculated value of f_c. Note the frequency value that causes the **minimum** amount of output voltage. Record this frequency on the Results Sheet.

4. Set the frequency of the function generator to the value of f_1 that you calculated in Step 1. Measure the values of f_1 and f_2 as you did in Step 4 of Part 1. Record the findings in Part 2 of the Results Sheet.

5. Calculate the dB voltage gain for each of the measurements in Table 3-2. Plot this output data on the semilog graph shown as Figure 3-2 on the Results Sheet. Also include the points for the measured values of f_c, f_1, and f_2 from Steps 3 and 4.

Click the exit button when you are ready to leave this project.

Name _____

Results Sheet

Project 3

Part 1   **Series *LC* Bandpass Filter**

Step 1

$L =$ _____ $C =$ _____

Calculated $f_c =$ _____

Calculated $f_1 =$ _____   Calculated $f_2 =$ _____

Step 3

Measured $f_c =$ _____

Step 4

Measured $f_1 =$ _____   Measured $f_2 =$ _____

Questions

1. What is the passband of the response curve for the circuit?

2. What is the Q of the circuit?

3. What are the approximate dB/decade slopes for the response curve?

f MHz	v_o V_{p-p}	dB Response
1.0		
3.0		
5.0		
7.0		
9.0		
10.0		
30.0		
50.0		
70.0		
90.0		
100.0		

Table 3-1

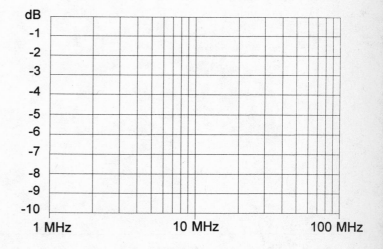

Figure 3-1

Part 2   Series *LC* Band-reject Filter

Step 1

$L =$ _____ $C =$ _____

Calculated $f_c =$ _____

Calculated $f_1 =$ _____   Calculated $f_2 =$ _____

Step 3

Measured $f_c =$ _____

Step 4

Measured $f_1 =$ _____   Measured $f_2 =$ _____

f kHz	v_o V_{p-p}	dB Response
2.5		
4.5		
6.5		
8.5		
15.0		
35.0		
55.0		
75.0		
95.0		
250.0		

Table 3-2

Questions

1. What is the bandwidth of the response curve for Part 2?

2. What is the Q of the circuit?

3. What are the approximate dB/decade slopes for the response curve?

Figure 3-2

Critical Thinking for Project 3

1. Explain why you adjust the frequency for a **maximum** output when finding the value of f_c for a bandpass circuit, and explain why you adjust the frequency for a **minimum** output when finding the value of f_c for a band-reject circuit.

2. Explain why the equations for an *LC* filter are identical to the equations for the resonant frequency of an *LC* circuit.

3. Explain why the circuit for Part 1 would respond as a high-pass filter circuit if the inductor were shorted. How would the circuit respond if the inductor is okay, but the capacitor is shorted?

4. The circuit for Part 2 uses a parallel *LC* circuit as a band-reject filter. Sketch an *LC* band-reject filter that uses a series *LC* circuit.

12

Project 4

Passive Filters II

A Hands-On Project

In this project you will construct passive *LC* filter circuits and determine their frequency response characteristics. In each part of the project you will:

- Construct the circuit.
- Calculate the center frequency.
- Gather data for the response curve.
- Plot the response curve on semilog graph.
- Determine the center frequency and cut-off frequencies from the curve.
- Confirm the actual center and cut-off frequencies by direct measurement.

Components and Supplies

1	Resistor, 820 Ω
2	Capacitor, 10 nF
2	Inductor, 1 mH

Preparation

Read Frenzel, *Principles of Electronic Communication Systems*, Section 2-3.

Complete the work for Prep Project 3.

Equipment

1	Function generator
1	Dual-trace oscilloscope
1	Frequency counter (optional)

The frequency counter is optional because it is possible to use the oscilloscope to determine the circuit's operating frequency.

Lab Procedure

Part 1   Series *LC* Bandpass Filter

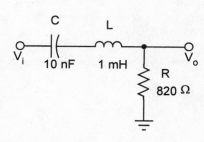

Figure 4-1

1. Construct the circuit of Figure 4-1. Connect the function generator, circuit, and oscilloscope as shown for Project 2.

2. Set the function generator to produce a sinusoidal waveform 10 $V_{p\text{-}p}$ at v_i.

3. For each of the entries in Table 4-1, set the function generator for the given frequency, double-check the value of v_i, and record the peak-to-peak value of v_o.

---

4. Calculate the center frequency f_c according to the values of L and C given in Figure 4-1. Record your results in Part 1 of the Results Sheet.

5. Set the frequency of the function generator to the calculated value of f_c, then carefully tune the applied frequency to obtain the maximum level of v_o. Record this measured center frequency on the Results Sheet.

6. Convert all values for v_o in Table 4-1 to their corresponding dB level. Plot the results on the two-cycle semilog graph of Figure 4-3 on the Results Sheet.

7. Estimate the values of f_1 and f_2 from the graph, and record your values on the Results Sheet.

8. Make sure v_i is still adjusted to 10 V_{p-p}, and determine the measured value of f_c by adjusting the frequency of the function generator to obtain the maximum value of v_o. Determine the actual value of f_1 by adjusting the frequency of the function generator below f_c at frequency $0.707 \times v_i$, and determine the actual value of f_2 above f_c at frequency $0.707 \times v_i$. Record your findings for these measured values on the Results Sheet.

Part 2 Shunt LC Band-reject Filter

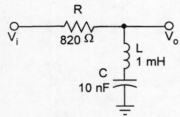

Figure 4-2

1. Construct the circuit of Figure 4-2, and connect the function generator and oscilloscope as described for Part 1 of this project.

2. For each of the entries in Table 4-2, set the function generator for the given frequency, double-check the 10 V_{p-p} value of v_i, and record the peak-to-peak value of v_o.

3. Calculate the center notch frequency f_{notch} according to the values of L and C given in Figure 4-2. Record your result on the Results Sheet.

4. Set the frequency of the function generator to the calculated value of f_{notch}, then fine-tune the applied frequency to obtain the minimum level of v_o. Record this measured center frequency and the maximum level of output voltage on the Results Sheet.

5. Convert all values for v_o in Table 4-2 to their corresponding dB level. Plot the results on the two-cycle semilog graph of Figure 4-4. **Note:** Plot attenuation values that are beyond -10 dB as -10 dB.

6. Estimate the values of f_1 and f_2 from the graph, and record your values on the Results Sheet.

7. Make sure v_i is still adjusted to 10 V_{p-p}, and determine the measured value of f_{notch} by adjusting the frequency of the function generator to obtain the minimum value of v_o. Record your finding on the Results Sheet.

8. Determine the actual value of f_1 by adjusting the frequency of the function generator below f_{notch} for the -3 dB value of v_o. Likewise, determine the actual value of f_2 by determining the -3 dB output at a frequency above f_{notch}. Record your findings for the measured values of the cut-off frequencies on the Results Sheet.

Results Sheet

Project 4

Part 1   Series LC Bandpass Filter

f kHz	v_o Vp-p	v_o/v_i	20 log(v_o/v_i) dB
1.5			
3			
6			
9			
12			
15			
30			
60			
90			
120			
150			
32.0			

Table 4-1

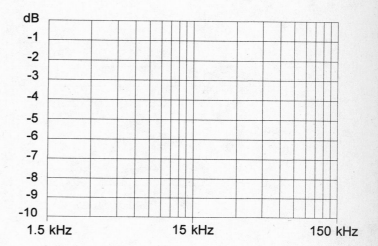

Figure 4-3

Step 4

    Calculated f_c = _____

Step 5

    Measured f_c = _____

Step 7

    Estimated f_1 = _____

    Estimated f_2 = _____

Step 8

    Measured f_1 = _____

    Measured f_2 = _____

Questions

1. What is the Q of the circuit at f_c?

2. What is the bandwidth of the circuit?

3. Describe any difference you see between calculated and measured values of f_1, f_2, and f_c. Explain your findings.

Part 2   Shunt *LC* Band-reject Filter

f kHz	v_o Vp-p	v_o/v_i	20 log(v_o/v_i) dB
1.5			
3			
6			
9			
12			
15			
30			
60			
90			
120			
150			
32.0			

Table 4-2

Step 3

     Calculated f_{notch} = _____

Step 4

     Measured f_{notch} = _____

     Measured $v_{o(min)}$ = _____

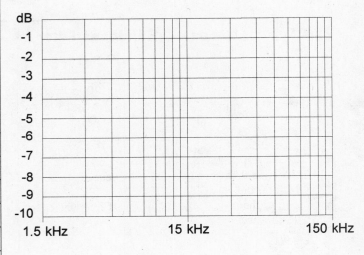

Figure 4-3

Step 6

     Estimated f_1 = _____

     Estimated f_2 = _____

Step 7

     Measured f_1 = _____

     Measured f_2 = _____

Questions

1. What is the Q of the circuit at f_{notch}?

2. What is the bandwidth of the circuit?

Critical Thinking for Project 4

1. Explain how the shunt resistor at the output of the filter circuit in Figure 4-2 affects the values of f_1, f_2, and f_c.

2. Describe the differences between the measured values of f_1 and f_2 for the filters you studied in Parts 1 and 2 of this project. Explain the differences you find.

3. If the inductor in Figure 4-1 is shorted, the circuit will behave as a different kind of filter circuit. Describe the kind of filter effect that would occur.

4. If the inductor in Figure 4-2 is open, the circuit will behave as a different kind of filter circuit. Describe the kind of filter effect that would occur.

Project 5

Active Filters

A Prep Project

This project is a computer simulation of laboratory tests for determining the frequency response of active bandpass and band-reject filters. In each case you will:

- Calculate the center frequency.
- Gather data for the response curve.
- Plot the response curve on a semilog graph.
- Determine the center frequency and cut-off frequencies from the curve.
- Confirm the actual center and cut-off frequencies by direct measurement.

Preparation

Read Frenzel, *Principles of Electronic Communication Systems*, Section 2-3.

Setup Procedure

1. Select Prep Projects from the Project menu.

2. Select Project 5 Active Filters from the list of Prep Projects.

Lab Procedure

For both parts of this project you will be using the simulated function generator and ac voltmeter. The function generator supplies the required input waveform. The ac voltmeter provides a convenient means for monitoring the output voltage level.

The block diagrams indicate how the equipment and circuits are interconnected. There are no schematic diagrams shown for this project because there are so many practical ways to go about performing the work of active filters. The data you gather here can represent many of the active bandpass and band-reject filters described in your textbook and in the hands-on version of Project 5.

Part 1   Active Bandpass Filter

Check the screen's project bar to confirm you are working with Part 1 of this project.

1. Adjust the amplitude of the function generator for 10.0 V_{p-p}. Adjust the frequency of the function generator to each of the values shown in Table 5-1 on the Results Sheet. In each case, record the value of v_o as shown on the ac voltmeter.

2. Plot this output data from Step 1 on the linear graph shown in Figure 5-1 on the Results Sheet.

3. Estimate the value of f_c from the curve in Figure 5-1 and adjust the frequency of the function generator to that value. While observing the output on the ac voltmeter, adjust the frequency slightly above and below your estimated value of f_c until you find a frequency that causes the **maximum** amount of output voltage. Record this frequency on the Results Sheet.

4. Set the frequency of the function generator to your estimated value f_1. While observing the output on the ac voltmeter, adjust the frequency slightly above and below the calculated value until the output voltage equals 0.707 times the maximum output voltage. Record this measured value of f_1 on the Results Sheet. Repeat the adjustments and measurements for the upper cut-off point, f_2.

Click the browse button to go to Part 2 of this project.

Part 2   Series *LC* Band-reject Filter

Check the screen's project bar to confirm you are working with Part 2 of this project.

1. Adjust the amplitude of the function generator for 10.0 V_{p-p}. Adjust the frequency of the function generator to each of the values shown in Table 5-2 on the Results Sheet. In each case, record the value of v_o as shown on the ac voltmeter.

2. Plot the data from Step 1 on the linear graph shown in Figure 5-2 on the Results Sheet.

3. As in the previous part of this project, estimate the values of f_c, f_1, and f_2 from the curve in Figure 5-2. Then adjust the frequency of the function generator to determine the precise values. (Remember to adjust for a **minimum** voltage value when verifying the value of f_c for a band-reject filter circuit.) Record your findings on the Results Sheet.

Click the exit button when you are ready to leave this project.

Name _____

Results Sheet

Project **5**

Part 1   Active Bandpass Filter

Step 3

     Measured f_c = _____

Step 4

     Measured f_1 = _____   Measured f_2 = _____

Questions

1. What is the voltage gain (in v_o / v_i) of this circuit at the true center frequency?

2. What is the passband of the response curve?

3. What is the value of Q for this circuit?

f kHz	v_o Vp-p	dB Response
400		
420		
440		
460		
480		
500		
520		
540		
560		
580		
600		

Table 5-1

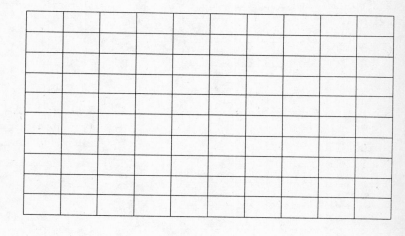

Frequency In (kHz)

Figure 5-1

Part 2 Series *LC* Band-Reject Filter

Step 3

Measured f_c = _____

Measured f_1 = _____   Measured f_2 = _____

Questions

1. From the data available in Figure 5-2, what is the maximum voltage gain (in v_o/v_i) for this circuit?

2. What is the bandwidth of the response curve in Figure 5-2?

3. What is the value of Q for this circuit? Show how you arrived at your answer.

f kHz	v_o $V_{p\text{-}p}$	dB Response
400		
420		
440		
460		
480		
500		
520		
540		
560		
580		
600		

Table 5-2

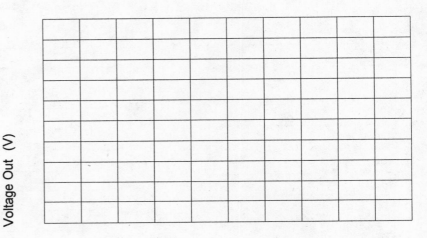

Frequency In (kHz)

Figure 5-2

Critical Thinking for Project 5

1. Name at least three advantages of active filters over their passive counterparts.

2. Describe the effect that decreasing the values of capacitors will most likely have upon the center frequency of an active bandpass filter circuit.

3. Sketch the circuit for a twin-T version of the active bandpass filter in Part 1.

4. Sketch the circuit for a twin-T version of the active notch filter in Part 2.

Project 6

Active Filters

A Hands-On Project

In this project you will construct the circuit and determine the frequency response characteristics of active *RC* band-pass and band-reject circuits. For each circuit you will:

- Construct the circuit.
- Calculate the center frequency.
- Gather data for the response curve.
- Plot the response curve on semilog graph.
- Determine the center frequency and cut-off frequencies from the curve.
- Confirm the actual center and cut-off frequencies by direct measurement.

Preparation

Read Frenzel, *Principles of Electronic Communication Systems*, Section 2-3.

Complete the work for Prep Project 5.

Components and Supplies

1	Resistor, 120 Ω
1	Resistor, 180 Ω
1	Resistor, 470 Ω
1	Resistor, 2.7 kΩ
1	Resistor, 3.3 kΩ
1	Resistor, 22 kΩ
1	Resistor, 330 kΩ
2	Capacitor, 1 nF
1	741 Op amp

Equipment

1	Power supply
1	Function generator
1	Dual-trace oscilloscope
1	Frequency counter

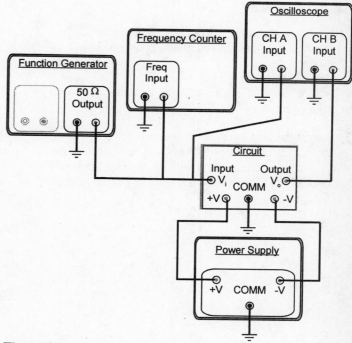

Figure 6-1

Lab Procedure

Part 1   Active Bandpass Filter

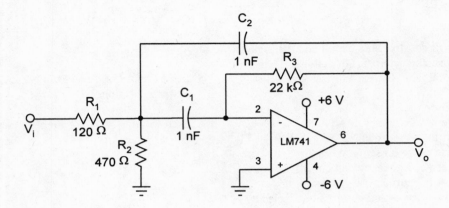

Figure 6-2

1. Construct the circuit of Figure 6-2, and connect the function generator and oscilloscope as shown in Figure 6-1.

2. Set the function generator for a sinusoidal waveform of 100 mV$_{p-p}$ at v_i.

3. Adjust the frequency of the function generator to obtain the maximum level of v_o. This will be the center frequency of the response waveform. Record the center frequency and the maximum level of output voltage.

4. Using the results in Step 3, calculate and record the value of v_o that should be present when the input frequency is at the lower and upper cut-off frequencies (f_1 and f_2.)

5. Adjust the frequency generator to obtain the values of v_o for f_1 and f_2 that you calculated in Step 4. Record these frequencies as the measured cut-off frequencies for this circuit.

6. Use the measured values you have determined for $f_1, f_c,$ and f_2 to sketch an approximation of the response curve for this circuit. Use the semilog graph in Figure 6-4.

Part 2   Active Band-reject Filter

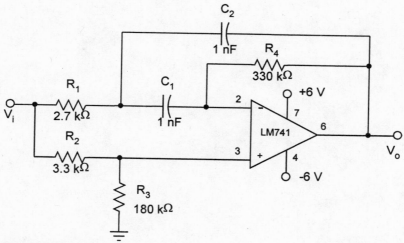

Figure 6-3

1. Construct the circuit shown in Figure 6-3, and connect the function generator and oscilloscope as shown in Figure 6-1.

2. Adjust the function generator for a 500 mV$_{(p-p)}$ output, and scan the frequency bands to determine f_{notch}. Record the minimum output voltage level and f_{notch} frequency on the Results Sheet. Also determine the maximum output voltage ($v_{o(max)}$) by tuning the function generator far from the value of f_{notch}. Record this value as well.

3. Calculate and record the half-power voltage level for this circuit as $0.707 \times v_{o(max)}$.

4. Adjust the input frequency to obtain the half-power voltage output level at points below and above f_{notch}. These are the cut-off frequencies (f_1 and f_2) for the circuit. Record the values on the Results Sheet.

5. Adjust the output of the frequency generator to each of the frequencies listed in Table 6-1. In each case, make sure $v_i = 500$ mV, and record the value of v_o.

6. Calculate the ratio $v_o/v_{o(max)}$ for each measurement, then convert the result to dB gain. Plot the results on the semilog graph in Figure 6-4. Also plot the values you determined in Steps 2 and 4.

Experimental Notes and Calculations

Name _____

Results Sheet

Project 6

Part 1   Active Bandpass Filter

Step 3

$f_c =$ _____ v_o at $f_{co} =$ _____

Step 4

Calculated v_o for $f_1 =$ _____

Calculated v_o for $f_2 =$ _____

Step 5

Measured value of $f_1 =$ _____

Measured value of $f_2 =$ _____

Questions

1. What is the voltage gain of the circuit at resonance?

2. What are the bandwidth and Q of the circuit?

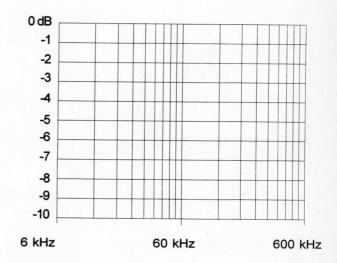

Figure 6-4

Part 2   Active Band-reject Filter

Step 2

f_{notch} = _____ $v_{o(min)}$ = _____

$v_{o(max)}$ = _____

Step 3

v_o at half-power points = _____

Step 6

f_1 = _____ f_2 = _____

f kHz	v_o $V_{p\text{-}p}$	v_o/v_i	20 log(v_o/v_i) dB
1.0			
2.0			
3.0			
4.0			
5.0			
6.0			
7.0			
8.0			
9.0			
10.0			

Table 6-1

Questions

1. What are the bandwidth and Q of the circuit?

2. What is the voltage gain (v_o/v_i) at f_{notch} when the input frequency is less than 1 kHz?

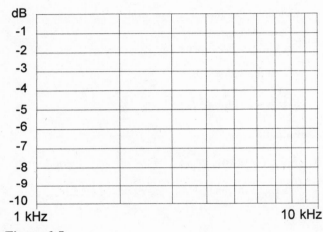

Figure 6-5

Critical Thinking for Project 6

1. Name the components in Figure 6-2 that most likely determine the value of the circuit's resonant frequency.

2. Describe the differences in components and connections between the two circuits in this project. Explain how these few differences account for a great difference in their operating characteristics.

3. Explain how the bandwidth of active filters (such as those in this project) can be decreased.

Project 7

Ceramic Filters

A Prep Project

This project is a computer simulation of laboratory tests for determining the frequency response of ceramic filters. During the course of this work, you will:

- Gather data for the response curve.
- Plot the response curve.
- Determine the relative amplitude and frequency of multiple response peaks.
- Determine the cut-off frequencies from the curve.
- Confirm the actual center and cut-off frequencies by direct measurement.
- Determine the bandwidth of the circuit.

Preparation

Read Frenzel, *Principles of Electronic Communication Systems*, Section 2-3.

Setup Procedure

1. Select Prep Projects from the Project menu.

2. Select Project 7 Ceramic Filters from the list of Prep Projects.

Lab Procedure

For both parts of this project you will be using the simulated function generator and ac voltmeter. The function generator supplies the required input waveform. The ac voltmeter provides a convenient means for monitoring the output voltage level.

The schematic and block diagrams are identical for both parts of the project. The only difference between Part 1 and Part 2 are the center frequency and bandwidth of the ceramic filters.

Part 1   455 kHz Filter

Check the project bar across the top of the screen to confirm you are working with Part 1 of this project.

1. Set the output of the function generator to 10 V. For each entry in Table 7-1 on the Results Sheet, set the function generator to the specified frequency and record the resulting peak-to-peak value of v_o.

2. Determine the maximum value of v_o from the data in Table 7-1. Use this value as the 0 dB level for converting all other values for v_o to their corresponding dB level.

3. Plot the values for v_o on the linear graph in Figure 7-1. The response curve should have three peaks. From the data you have gathered, cite the frequency and amplitude of each peak. Use the spaces provided for these values on your Results Sheet.

4. From your response curve (see Figure 7-1), determine the upper and lower cut-off frequencies. Record your answers on the Results Sheet.

Click the browse button to go to Part 2 of this project.

Part 2   50 MHz Filter

Check the screen's project bar to confirm you are working with Part 2 of this project.

1. For each of the entries in Table 7-2 on the Results Sheet, set the function generator to the specified frequency and record the resulting peak-to-peak value of v_o.

2. Determine the maximum value of v_o from the data in Table 7-2. Use this value as the 0 dB level for converting all other values for v_o in Table 7-2 to their corresponding dB level.

3. Plot the results on the linear graph in Figure 7-2. The response curve should have five peaks. Indicate on the graph the location of each peak. Estimate the frequency and amplitude of each peak. Verify your estimates by actually tuning the frequency source back and forth across each peak. Record the findings on the Results Sheet.

4. From the response curve in Figure 7-2, determine the upper and lower cut-off frequencies of the filter. Record your answers on the Results Sheet.

Click the exit button when you are ready to leave this project.

Results Sheet

Project 7

Part 1   455 kHz Filter

f kHz	v_o Vp-p	dB Response
420		
430		
440		
445		
450		
455		
460		
465		
470		
480		
490		

Table 7-1

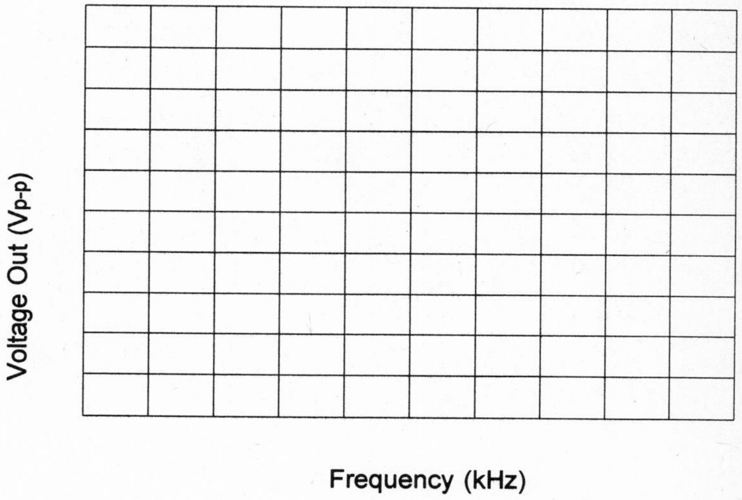

Frequency (kHz)

Figure 7-1

Step 3

　　Lower frequency peak:

　　　　$f =$ _____ $v =$ _____

　　Middle frequency peak:

　　　　$f =$ _____ $v =$ _____

　　Upper frequency peak:

　　　　$f =$ _____ $v =$ _____

Step 4

　　Lower cut-off frequency $f_1 =$ _____

　　Upper cut-off frequency $f_2 =$ _____

Questions

1. What is the center frequency for the response curve in Figure 7-1? Explain your answer.

2. What is the passband of the response curve in Figure 7-1?

3. What is the Q of the response curve in Figure 7-1?

Part 2   50 MHz Filter

f MHz	v_o V_{p-p}	dB Response
38		
40		
42		
44		
46		
48		
50		
52		
54		
56		
58		
60		
62		

Table 7-2

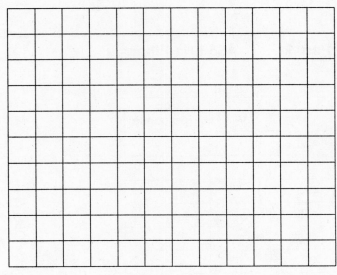

Figure 7-2

Step 3

First peak (lowest frequency):

$f =$ _____ $v =$ _____

Second peak:

$f =$ _____ $v =$ _____

Third peak:

$f =$ _____ $v =$ _____

Fourth peak:

$f =$ _____ $v =$ _____

Fifth peak (highest frequency):

$f =$ _____ $v =$ _____

Step 4

Lower cut-off frequency $f_1 =$ _____

Upper cut-off frequency $f_2 =$ _____

Questions

1. What is the center frequency for the response curve in Figure 7-2? Explain your answer.

2. What is the passband of the response curve in Figure 7-2?

3. What is the Q of the response curve in Figure 7-2?

Critical Thinking for Project 7

1. Describe how the value of Q for a typical ceramic filter compares with the Q for a typical quartz crystal.

2. Explain why a ceramic filter, used alone as in this project, is **not** considered an active filter.

3. Describe how a ceramic filter might be connected as a band-reject filter.

Experimental Notes and Calculations

Project 8

Ceramic Filters

A Hands-On Project

In this project you will determine the frequency response characteristics of a 455 kHz ceramic filter. The work requires you to:

- Construct the circuit.
- Gather data for the response curve.
- Plot the response curve .
- Determine the relative amplitude and frequency of multiple response peaks.
- Determine the cut-off frequencies from the curve.
- Confirm the actual peak and cut-off frequencies by direct measurement.

Preparation

Read Frenzel, *Principles of Electronic Communication Systems*, Section 2-3.

Complete the work for Prep Project 7.

Ceramic filters often exhibit more than one response peak, the number depending on the manufacturer and the design specifications for the device.

Components and Supplies

2   Resistor, 10 kΩ
1   455 kHz ceramic filter

Equipment

1   Function generator
1   Dual-trace oscilloscope
1   Frequency counter

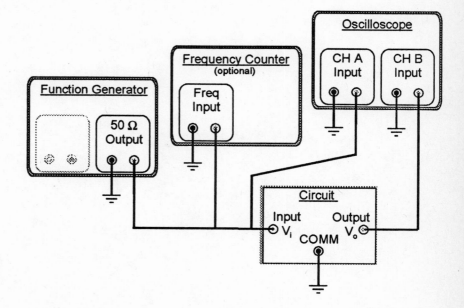

Figure 8-1

Lab Procedure

1. Construct the circuit of Figure 8-1. Connect the function generator, oscilloscope, and frequency counter as shown in Figure 8-1.

2. Adjust the function generator to obtain a sinusoidal waveform of 10 V_{p-p} at v_i.

3. Set the frequency of the function generator to 355 kHz, then increase the applied frequency while noting the level of v_o. Look for more than one response peak at v_o. When you find a peak, record the frequency and voltage v_o on the Results Sheet.

4. Determine the maximum value of v_o from your data in Table 8-1. Use this value as the 0 dB level for determine the value of v_o you will find at the -3 dB cut-off frequencies. Record these values on the Results Sheet.

5. Plot the response curve for this device on the graph in Figure 8-3.

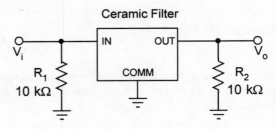

Ceramic Filter

Figure 8-2

Results Sheet

Project 8

Step 3

Peak #1

    freq = _____ v_o = _____

Peak #2

    freq = _____ v_o = _____

Peak #3

    freq = _____ v_o = _____

Step 5

    Calculated value of v_o at the
    cut-off frequencies = _____

Step 6

    Measured f_1 = _____   Measured f_2 = _____

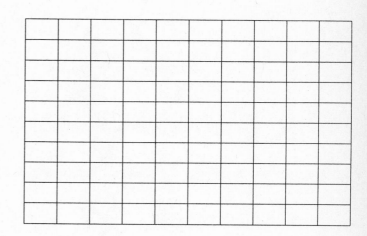

Figure 8-3

Questions

1. How many peaks did you find? What are their frequencies?

2. What is the bandwidth of the circuit?

3. What is the Q of the circuit?

Critical Thinking for Project 8

1. Cite some practical advantages of using a ceramic filter as opposed to *LC* or *RC* bandpass filters.

2. What are some disadvantages of using ceramic filters as opposed to *LC* or *RC* bandpass circuits?

Experimental Notes and Calculations

Project 9

Sweep Generators

An Extended Project

This project uses a simulated frequency sweep generator and an oscilloscope display to determine the response curve for the most common types of filter circuits. You will:

- Calculate the cut-off frequencies and center frequencies of various types of filter circuits.
- Determine the cut-off frequencies from sweep-generator displays of low- and high-pass filters.
- Determine the cut-off and center frequencies from sweep-generator displays of bandpass and band-reject filters.

Preparation

Read Frenzel, *Principles of Electronic Communication Systems*, Section 2-3.

Setup Procedure

1. Select Demo Projects from the Project menu.

2. Select Project 9 Sweep Generators from the list of Extended Projects.

Lab Procedure

The simulated sweep generator includes the controls required for working with filter circuits. The block diagram shows that outputs of the sweep generator include the sweep frequency (applied at the input of the circuit under test) and a horizontal sawtooth waveform (applied to the horizontal input of the oscilloscope). The output of the circuit under test is applied to the vertical input of the oscilloscope. With this arrangement, the oscilloscope directly displays the amplitude versus frequency response curves you have produced in some of the previous projects.

Part 1   Low-Pass Filter

Check the project bar across the top of the screen to confirm you are working with Part 1 of this project.

1. Set the amplitude of the sweep generator to 12.6 V. Use this setting for all parts of the project.

2. Set the sweep generator frequency to 0.1 MHz. This is the frequency where the sweep on the oscilloscope display begins.

3. Set the sweep range to 999.9 MHz. This determines the frequency range of the horizontal sweep on the oscilloscope covers.

With the frequency set to 0.1 MHz and the sweep range to 999.9 MHz, the sweep on the oscilloscope begins at 0.1 MHz and ends at 0.1 MHz + 999.9 MHz, or 1000 MHz.

Also notice that there are ten horizontal divisions on the oscilloscope display. The sweep range is set for 1000 MHz. This means the horizontal scaling on the oscilloscope is 1000/10, or 100 MHz/div (megahertz per division).

4. Sketch the oscilloscope display on the grid provided as Figure 9-1 on the Results Sheet. Mark the cut-off point on your drawing, and record the cut-off frequency.

Hint: This waveform has an amplitude that covers three vertical divisions. Therefore the cut-off amplitude is 0.707×3 divisions from the baseline of the waveform.

Click the browse button to go to Part 2 of this project.

Part 2   High-Pass Filter

Check the screen's project bar to confirm you are working with Part 2 of this project.

1. Make sure the amplitude is still set for 12.6 V.

2. Set the frequency to 0.1 MHz and set the sweep range to 999.9 MHz. Determine the frequency at the left side of the oscilloscope display, the frequency at the right side of the display, and the horizontal scaling. Record your values on the Results Sheet.

3. Sketch the oscilloscope display on the grid provided as Figure 9-2 on the Results Sheet. Mark the point of the cut-off frequency on your drawing and record the value on the Results Sheet.

   Click the browse button to go to Part 3.

Part 3   Bandpass Filter

Check the screen's project bar to confirm you are working with Part 3 of this project.

1. Make sure the amplitude is still set for 12.6 V.

2. Set the frequency to 312.0 MHz and set the sweep range to 375.0 MHz. Determine the frequency at the left side of the oscilloscope display, the frequency at the right side of the display, and the horizontal scaling. Record your values on the Results Sheet.

3. Sketch the oscilloscope display on the grid provided as Figure 9-3 on the Results Sheet. On your drawing, mark the points for the center frequency, the upper cut-off frequency, and the lower cut-off frequency. Also record these values on the Results Sheet.

   Click the browse button to go to Part 4.

Part 4   Band-Reject Filter

Check the screen's project bar to confirm you are working with Part 4 of this project.

1. Make sure the amplitude is still set for 12.6 V.

2. Set the frequency to 375.0 MHz and set the sweep range to 250.0 MHz. Determine the frequency at the left side of the oscilloscope display, the frequency at the right side of the display, and the horizontal scaling. Record your values on the Results Sheet.

3. Sketch the oscilloscope display on the grid provided as Figure 9-4 on the Results Sheet. On your drawing, mark the points for the notch frequency, the upper cut-off frequency, and the lower cut-off frequency. Also record these values on the Results Sheet.

   Click the browse button to go to Part 5.

Part 5   Ceramic Filter

Check the screen's project bar to confirm you are working with Part 5 of this project.

1. Make sure the amplitude is still set for 12.6 V.

2. Set the frequency to 250.0 MHz and set the sweep range to 500.0 MHz. Determine the frequency at the left side of the oscilloscope display, the frequency at the right side of the display, and the horizontal scaling. Record your values on the Results Sheet.

3. Sketch the oscilloscope display on the grid provided as Figure 9-5 on the Results Sheet. Determine all peak frequencies and the upper and lower cut-off frequencies. Also record these values on the Results Sheet.

Click the exit button when you are ready to leave this project.

Experimental Notes and Calculations

Results Sheet

Project 9

Part 1   Low-Pass Filter

Step 4

     Estimated f_{co} = _____

Questions

1. Assuming the amplitude of the output waveform is 12.6 V, what is the voltage at the −3 dB level?

2. How many horizontal divisions of this display make up a decade?

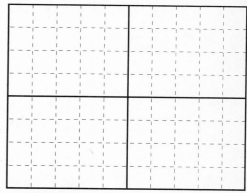

Figure 9-1

Part 2   High-Pass Filter

Step 2

     Frequency at start of the sweep = _____

     Frequency at the end of the sweep = _____

     Horizontal scaling of the sweep = _____

Step 3

     Estimated f_{co} = _____

Questions

1. Assuming the amplitude of the output waveform is 12.6 V, what is the voltage at the −3 dB level?

2. This waveform is taken from the output of a passive *RC* filter. If the value of *R* is known to be 47 kΩ, what is the value of the capacitor?

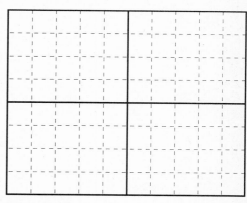

Figure 9-2

Part 3   Bandpass Filter

Step 2

Frequency at start of the sweep = _____

Frequency at the end of the sweep = _____

Horizontal scaling of the sweep = _____

Step 3

Measured f_c = _____

Measured f_1 = _____   Measured f_2 = _____

Questions

1. If the vertical scaling of the display is actually 1 V/div, what is the voltage at the −3 dB level on the waveform?

2. What is the bandwidth of the filter used in Part 3?

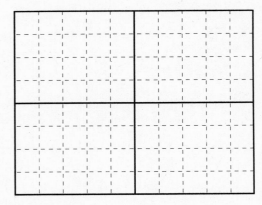

Figure 9-3

Part 4   Band-reject Filter

Step 2

Frequency at start of the sweep = _____

Frequency at the end of the sweep = _____

Horizontal scaling of the sweep = _____

Step 3

Estimated f_{notch} = _____

Estimated f_1 = _____   Estimated f_2 = _____

Questions

1. What is the bandwidth of the filter used in Part 4?

2. With a scaling of 3.2 V/div, what is the voltage at the peak? the half-power points?

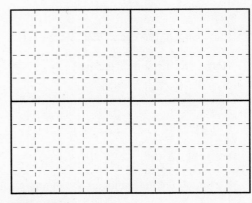

Figure 9-4

Part 5   Ceramic Filter

Step 2

Frequency at start of the sweep = _____

Frequency at the end of the sweep = _____

Horizontal scaling of the sweep = _____

Step 3

Estimated f_c = _____

Estimated f_1 = _____   Estimated f_2 = _____

Estimated frequencies of the two lower peaks = _____ and _____

Estimated frequencies of the two upper peaks = _____ and _____

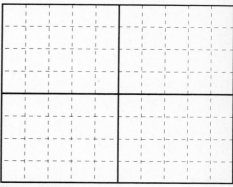

Figure 9-5

Questions

1. What is the bandwidth of the filter in Part 5?

2. If the scaling is 1.5 V/div, what is the peak voltage? the voltage at the −3 dB points?

Critical Thinking for Project 9

1. The horizontal axis of all the displays in this project is linear. What would be the advantage of using a display that had a semilog scale for its horizontal axis?

2. How do the settings for the frequency and sweep range affect the frequency and amplitude of the half-power points?

3. Explain how a sweep generator is basically a voltage-to-frequency converter that has a sawtooth waveform applied to it.

Experimental Notes and Calculations

Project 10

Fourier Theory

An Extended Project

This project is a computer simulation of laboratory tests for determining the harmonic frequencies of sinusoidal, square, and triangular waveforms that can pass through a bandpass filter circuit. For each of these waveforms, you will:

- Determine the first through fifth harmonic frequency content.
- Verify the harmonic content by manually tuning an output filter for peak voltage levels.

Preparation

Read Frenzel, *Principles of Electronic Communication Systems*, Section 2-5.

Setup Procedure

1. Select Extended Projects from the Project menu.

2. Select Project 10 Fourier Theory from the list of Extended Projects.

Lab Procedure

Note the instruments and circuits used in this project. In each part of the project, you will use a function generator to apply the specified waveform shape, amplitude, and frequency to a bandpass filter circuit. You will tune the bandpass filter and note the frequencies that can pass through it. These frequencies are the harmonic content of the waveform. You will use an analog voltmeter to read the amplitude of the waveform from the tuned filter.

Part 1   Sinusoidal Waveforms

Check the project bar across the top of the screen to confirm you are working with Part 1 of this project.

1. Calculate the frequencies for harmonics 1, 2, 3, 4, and 5 for a sinusoidal waveform of 100 kHz. List your results in the second column of Table 10-1 on the Results Sheet.

2. Set the amplitude of the function generator to its maximum output — 49.9 V.

3. Adjust the bandpass filter to each of the five harmonic frequencies you calculated. Determine the corresponding output voltage level for each harmonic, and record the values in the third column of Table 10-1.

Click the browse button to go to Part 2.

Part 2   Triangular Waveforms

Check the project bar across the top of the screen to confirm you are working with Part 2 of this project.

1. Calculate the frequencies for harmonics 1, 2, 3, 4, and 5 for a triangular waveform of 100 kHz. List your results in the first column of Table 10-2 of the Results Sheet.

2. Set the amplitude of the function generator to its maximum output.

3. Adjust the bandpass filter to each of the five harmonic frequencies you calculated. Determine the corresponding output voltage level for each harmonic, and record the values in the second column of Table 10-2.

Click the browse button to go to Part 3.

Part 3   Rectangular Waveforms

Check the project bar across the top of the screen to confirm you are working with Part 3 of this project.

1. Calculate the frequencies for harmonics 1, 2, 3, 4, and 5 for a rectangular waveform of 100 kHz. List your results in the first column of Table 10-3 on the Results Sheet.

2. Set the amplitude of the function generator to its maximum output.

3. Adjust the bandpass filter to each of the five harmonic frequencies you calculated. Determine the corresponding output voltage level for each harmonic, and record the values in the second column of Table 10-3.

Click the exit button when you are ready to leave this project.

Results Sheet

Project 10

Part 1   Sinusoidal Waveforms

Harmonic	Frequency (calculated)	v_o (measured)
First		
Second		
Third		
Fourth		
Fifth		

Table 10-1

Questions

1. Which harmonics are present and which are not?

2. If you assign 0 dB to the value of v_o at the fundamental frequency, what is the dB gain for each of the harmonics?

Part 2   Triangular Waveforms

Harmonic	Frequency (calculated)	v_o (measured)
First		
Second		
Third		
Fourth		
Fifth		

Table 10-2

Questions

1. Which harmonics are present and which are not?

2. If you assign 0 dB to the value of v_o at the fundamental frequency, what is the dB gain for each of the harmonics?

Part 3   Rectangular Waveforms

Harmonic	Frequency (calculated)	v_o (measured)
First		
Second		
Third		
Fourth		
Fifth		

Questions

1. Which harmonics are present and which are not?

2. If you assign 0 dB to the value of v_o at the fundamental frequency, what is the dB gain for each of the harmonics?

Table 10-3

Critical Thinking for Project 10

1. Write out the Fourier expressions for the triangular and rectangular waveforms. Circle the factors in the equations that indicate the presence of odd harmonics.

2. Waveforms do not usually have perfect shapes. Explain why you might expect to find some even harmonic content in the analysis of a rectangular waveform that is slightly rounded on the corners.

Project 11

Amplitude Modulation I

A Prep Project

In this project you will perform computer simulations of tests on a diode amplitude modulator circuit. You will observe and adjust the percent modulation of an AM envelope. You will:

- Sketch and identify the major components of an AM envelope.
- Adjust the AM output for designated levels of modulation.
- Determine the level of modulation from a given modulation waveform.

Preparation

Read Frenzel, *Principles of Electronic Communication Systems*, Section 4-2.

Setup Procedure

1. Select Demo Projects from the Project menu.

2. Select Project 11 Amplitude Modulation I from the list of Prep Projects.

Lab Procedure

This project uses simulated versions of a function generator, rf generator, and oscilloscope display. For the purposes of this project, the rf generator provides the carrier signal and the function generator supplies the audio modulating signal. The frequencies of the function generator and rf generator are not important for the work in this project, so they are fixed at 400 Hz and 1460 kHz, respectively.

The block diagram shows how these instruments are interconnected with the circuit. See your textbook for schematic diagrams of simple diode modulator circuits.

1. Set the output of the function generator for 0.0 V, and adjust the output of the rf generator for an output of 15.0 V. Record this voltage on the Results Sheet. This is the value of v_0 when there is no audio input.

Note: When the rf output is set to 15.0 V and the audio output from the function generator is set to 0.0 V, you can see that the oscilloscope is vertically scaled at 7.5 V/div. Use this value for determining other peak-to-peak values through this project.

2. Adjust the output of the function generator for an output of 15.0 V. Sketch the waveform at v_0, using the grid provided on the Results Sheet as Figure 11-1. Determine the percent modulation of this waveform and record the value on the Results Sheet.

3. Make sure the output of the rf generator is still set for 15.0 V, and then adjust the output of the function generator for an output of 7.5 V. Sketch the waveform at v_0, using the grid provided as Figure 11-2. Determine and record the percent of modulation on the Results Sheet.

4. Making sure that the output of the rf generator is still set for 15.0 V, adjust the function generator for an output of 3.5 V. Sketch the waveform at v_0, using the grid provided as Figure 11-3. Also determine and record the percent of modulation on the Results Sheet.

5. Adjust the output of the rf generator to 10.0 V. Adjust the output of the function generator in order to produce 75% modulation. Sketch the waveform on the grid provided as Figure 11-4. Read the following values from the oscilloscope display: V_{max}, V_{min}, $V_{max(p-p)}$, and $V_{min(p-p)}$.

6. Make sure the output of the rf generator is still at 10.0 V. Adjust the output of the function generator for 12.5 V. Sketch the waveform at v_0, using the grid provided as Figure 11-5. Also determine and record the percent of modulation on the Results Sheet.

Click the exit button when you are ready to leave this project.

Experimental Notes and Calculations

Results Sheet

Project 11

Step 1

v_o (audio = 0.0 V) = _____ $V_{p\text{-}p}$

Step 2

% modulation = _____

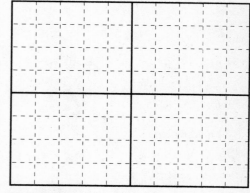

Figure 11-1

Step 3

% modulation = _____

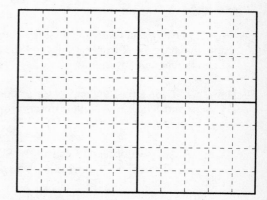

Figure 11-2

Step 4

% modulation = _____

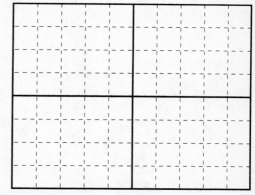

Figure 11-3

Step 5

$V_{max} =$ _____ $V_{min} =$ _____

$V_{max(p-p)} =$ _____ $V_{min(p-p)} =$ _____

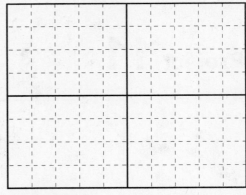

Figure 11-4

Step 6

% modulation = _____

Questions

1. What is the equation for calculating the percent of modulation, given the peak-to-peak values of the carrier signal and the modulating signal?

2. How should you adjust the amplitude settings on the rf generator and function generator in this project to produce 0% modulation?

3. Do any of the waveforms on this Results Sheet represent a condition of overmodulation? If so, which one?

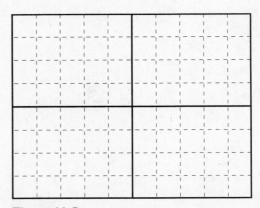

Figure 11-5

Critical Thinking for Project 11

1. Describe the special appearance of a modulation envelope for a signal that is modulated in excess of 100%.

2. Sketch the modulation envelopes for a signal that is being modulated with a waveform at 50% and at 100%.

Project 12

Amplitude Modulation I

A Hands-On Project

In this project you will construct and observe the operation of a simple diode amplitude modulator circuit. The purpose is to acquaint you with the most important features of an AM waveform. Through this work you will:

- Construct the circuit
- Sketch and identify the major components of an AM envelope.
- Adjust the AM output for designated levels of modulation.
- Determine the level of modulation from a given waveform.
- Sketch modulation waveforms for sinusoidal, triangular, and rectangular audio signals.

Components and Supplies

2	Resistor, 2.2 kΩ
1	Resistor, 100 kΩ
1	Capacitor, 1 μF
1	Capacitor, 10 nF
1	Capacitor, 100 nF
1	Inductor, 1 mH
1	NPN transistor, 2N3904 or equivalent

Equipment

1	Dual-trace oscilloscope
2	Function generator
1	Frequency counter (optional)
1	AM radio receiver

In this project, use one function generator as the audio source, and the second function generator as the carrier source.

Preparation

Read Frenzel, *Principles of Electronic Communication Systems*, Section 4-2.

Complete the work for Prep Project 11.

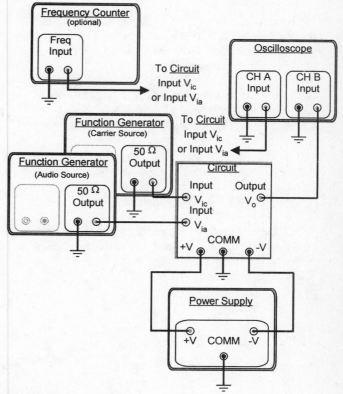

Figure 12-1

Lab Procedure

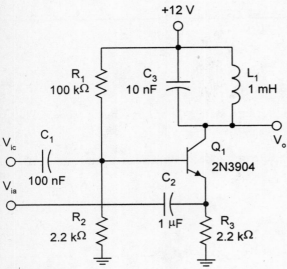

Figure 12-2

1. Construct the circuit shown in Figure 12-2. Connect the oscilloscope to output v_o, the carrier source (rf generator) to input v_{ic}, and the audio source (function generator) to input v_{ia} (see Figure 12-1).

2. Adjust the carrier source for a 1 V sinusoidal waveform at 50 kHz, and set the audio source for 0 V at 440 Hz. Slightly adjust the frequency of the carrier source for a peak output carrier waveform at v_o.

3. Adjust the audio source for a 500 mV output at 440 Hz. Adjust the sweep rate of the oscilloscope until you see a stable modulation envelope.

4. Adjust the amplitude of the audio source to obtain 100% modulation at v_o. Sketch about three cycles of the waveform in Figure 12-3 on the Results Sheet. Also indicate the value of $V_{max(p-p)}$ on the drawing.

5. Adjust the amplitude of the audio source to obtain 50% modulation. Sketch about three cycles of the waveform in Figure 12-4 on the Results Sheet. Also indicate the values of $V_{max(p-p)}$ and $V_{min(p-p)}$ on the drawing.

6. Readjust the amplitude of the audio source to obtain 100% modulation. While observing the modulation waveform on the oscilloscope, increase the amplitude of the audio source until you clearly see a waveform that indicates overmodulation. Sketch three cycles of the overmodulation waveform in Figure 12-6 on the Results Sheet. Also indicate the values of $v_{max(p-p)}$ and $v_{min(p-p)}$ on the drawing.

7. With the modulation still set at 100%, bring an AM radio receiver close to the circuit. Tune the radio receiver in the 1 MHz (1000 kHz) range until you hear the 440 Hz modulation signal. Tune the receiver for the loudest 440 Hz signal, and note the approximate carrier frequency as determined by reading the tuning dial or LCD display on the receiver. Record this frequency on the Results Sheet.

8. Reduce the level of modulation and note what happens to the sound level from the radio receiver. Note the relationship between percent modulation and the audio signal level from the receiver. Be prepared to describe this effect in the Questions section of the Results Sheet.

9. Switch the audio source to generate a triangular waveform at 1 MHz. Adjust the output level for 100% modulation as indicated by the waveform at v_o. Sketch three cycles of this modulation waveform in Figure 12-7. Also note the value of $V_{max(p-p)}$ on the drawing.

10. Switch the audio source to generate a rectangular waveform at 1 MHz and adjust the output level for 100% modulation. Sketch three cycles of this modulation waveform in Figure 12-8. Also note the value of $V_{max(p-p)}$ on the drawing.

Results Sheet

Project 12

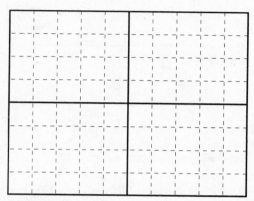

Figure 12-3

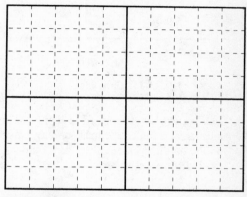

Figure 12-4

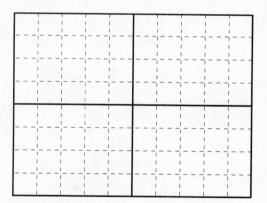

Figure 12-5

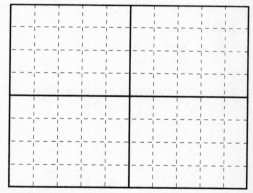

Figure 12-6

Step 7

    Receiver frequency = _____

Step 8

    Receiver frequency = _____

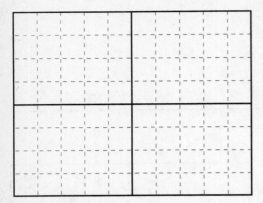

Figure 12-7

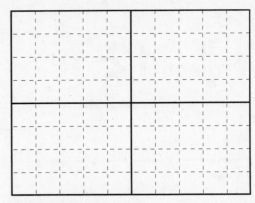

Figure 12-8

Questions

1. The voltage levels applied to inputs v_{ic} and v_{ia} often appear greater than the corresponding values that appear on the modulation envelope as v_c and v_m. How can these differences in amplitude be explained?

2. How would you describe the relationship between percent modulation and the audio signal level from the receiver as noted in Step 8 of this project?

3. What is the *main* purpose of the transistor in this AM modulator circuit?

Critical Thinking for Project 12

1. Explain why capacitor C_2 has a larger value than capacitor C_1.

2. Describe the purpose of the LC resonant circuit and explain why it is important for proper operation of an AM modulator circuit.

Project 13

Amplitude Modulation II

A Prep Project

This project simulates two different setups for reading the modulation level of an AM signal. You have already worked with one of the setups--the time-domain modulation display. This project introduces the trapezoidal display. You will:

- Compare time-domain and trapezoidal displays of typical AM signals.
- Determine the percent of modulation from oscilloscope displays.
- Determine the display that represents a desired level of amplitude modulation.

Preparation

Read Frenzel, *Principles of Electronic Communication Systems*, Section 4-2.

Setup Procedure

1. Select Prep Projects from the Project menu.

2. Select Project 13 Amplitude Modulation II from the list of Prep Projects.

Lab Procedure

This project uses simulated versions of a function generator, rf generator, and oscilloscope display. The rf generator provides the carrier signal, and the function generator supplies the audio modulating signal.

Connections to the oscilloscope are different for the time-domain and trapezoidal displays. Clicking the Mode button toggles the system between the two displays. Notice how the block diagrams for the two modes indicate the connections to the oscilloscope.

The formulas for calculating the percent of modulation for the time-domain and trapezoidal displays are essentially the same. The way the measurements are taken are different, however. Clicking the Calc button displays figures that indicate how you should determine the proper values from the oscilloscope.

1. Adjust the amplitude of the rf generator to 30.0 V and the amplitude of the function generator to 15.0 V. Set the project for the time-domain display, calculate the percent of modulation, and sketch the time-domain waveform in Figure 13-1.

2. Make sure the signal inputs are at the values used in Step 1. Set the project for the trapezoidal display, calculate the percent of modulation, and sketch the oscilloscope waveform in Figure 13-2.

3. Adjust the amplitude of the rf generator to 30.0 V and the amplitude of the function generator to 30.0 V. Set the project for the time-domain display, calculate the percent of modulation, and sketch the time-domain waveform in Figure 13-3.

4. Make sure the signal inputs are at the values used in the previous step. Set the project for the trapezoidal display, calculate the percent of modulation, and sketch the oscilloscope waveform in Figure 13-4.

5. Adjust the amplitude of the rf generator to 15.0 V and the amplitude of the function generator to 30.0 V. Set the project for the time-domain display, calculate the percent of modulation, and sketch the time-domain waveform in Figure 13-5.

6. Make sure the signal inputs are at the values used in the previous step. Set the project for the trapezoidal display, calculate the percent of modulation, and sketch the oscilloscope waveform in Figure 13-6.

7. Set the amplitude of the rf generator to 32.0 V. Calculate the amplitude of the audio signal (v_{ia}) required for 80% modulation. Record this value of v_{ia} on the Results Sheet and adjust the amplitude of the function generator to that value.

---

8. Set the project for the time-domain display, sketch the time-domain waveform in Figure 13-7, and determine the values for V_{min} and V_{max} from the display. Record these values on the Results Sheet.

9. Make sure the signal inputs are at the values used in Step 7. Set the project for the trapezoidal display, sketch the waveform in Figure 13-8, and determine the values for V_{min} and V_{max} from the display. Record these values on the Results Sheet.

Click the exit button when you are ready to leave this project.

Results Sheet

Project **13**

_____

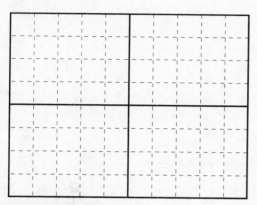

Figure 13-1

Step 1

    % mod = _____

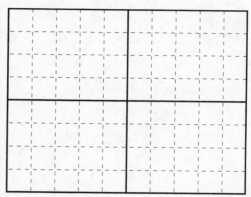

Figure 13-2

Step 2

    % mod = _____

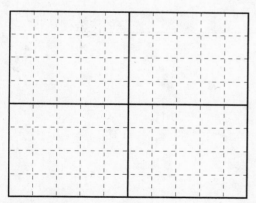

Figure 13-3

Step 3

    % mod = _____

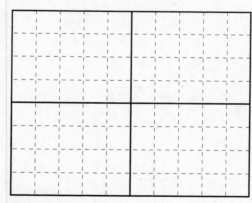

Figure 13-4

Step 4

    % mod = _____

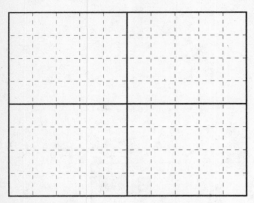

Figure 13-5

Step 5

   % mod = _____

Step 7

   v_{ia} = _____

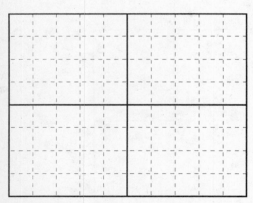

Figure 13-8

Step 9

   v_{ia} = _____

   V_{min} = _____ V_{max} = _____

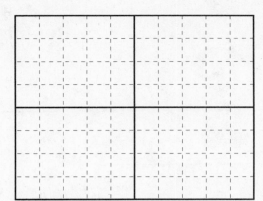

Figure 13-6

Step 6

   % mod = _____

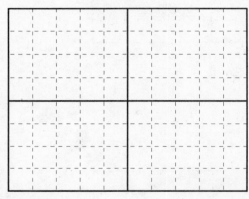

Figure 13-7

Step 8

   V_{min} = _____ V_{max} = _____

Questions

1. Which figures on this Results Sheet represent modulation levels less than 100%?

2. Which figures represent 100% modulation?

3. Which figures represent modulation levels greater than 100%?

Critical Thinking for Project 13

1. Give verbal descriptions of a trapezoidal waveform at (a) modulation less than 100%, (b) modulation at 100%, and (c) modulation greater than 100%.

2. Describe the direct effect, if any, that changes in the carrier frequency would have on the shape and size of a trapezoidal waveform.

3. This project simulation represents an ideal circuit. In real-life circuits, it is better to determine the modulation index from values of V_{max} and V_{min} on the oscilloscope than from values of v_{ia} and v_{ic} at the inputs. Explain why this is so.

Experimental Notes and Calculations

Project **14**

Amplitude Modulation II

A Hands-On Project

In this project you will demonstrate the use of an operational transconductance amplifier (OTA) as an amplitude modulator circuit. You will:

- Construct the circuit and null its output.
- Observe and adjust the percent modulation of AM envelope and trapezoidal waveforms.
- Adjust the carrier and audio inputs to achieve a designated level of modulation.
- Note the effects of overmodulation as determined by a trapezoidal waveform.

Components and Supplies

2	Resistor, 10 kΩ
1	Resistor, 47 kΩ
1	Resistor, 100 kΩ
1	Operational transconductance amplifier (OTA), CA3080

Equipment

1	Dual-voltage power supply
1	Dual-trace oscilloscope
2	Function generator
1	Frequency counter (optional)

In both parts of this project, one function generator serves as the audio source and the other as the carrier source.

Preparation

Read Frenzel, *Principles of Electronic Communication Systems*, Section 4-2.

Complete the work for Prep Project 13.

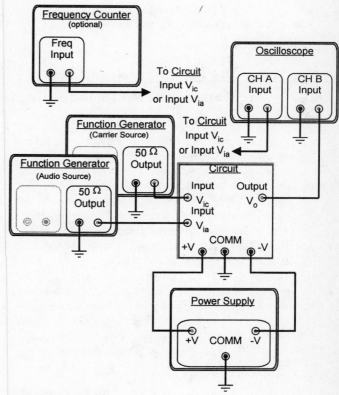

Figure 14-1

Lab Procedure

Part 1 Time-Domain Modulation Display

1. Construct the OTA amplitude modulator circuit shown in Figure 14-2. Connect the oscilloscope to v_o, the carrier source to v_{ic}, and the audio source to v_{ia} (see Figure 14-1).

2. Set the carrier source for a 1 $V_{p\text{-}p}$ sinusoidal waveform at 1 MHz, and adjust the audio source for a 1 $V_{p\text{-}p}$ sinusoidal waveform at 440 Hz. Adjust the sweep on the oscilloscope to show a stable image of the 440 Hz modulation envelope.

3. Adjust the amplitude of the audio source from 0 V to 1.5 V, noting the modulation response on the oscilloscope display. Determine the level of v_{ic} that causes 100% modulation. Record this value on the Results Sheet.

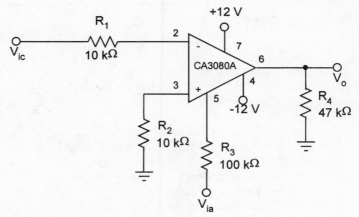

Figure 14-2

Part 2 Trapezoidal Modulation Display

This part of the project requires you to calculate the percent of modulation of an AM envelope from a trapezoidal display. The formula is the same as the formula for determing the percent of modulation from a time-domain AM display. For the trapezoidal display, however, the maximum voltage level refers to the amount of vertical deflection along the right side of the waveform (see the examples in Figure 14-4). The minimum voltage is taken as the amount of vertical deflection along the left side of the display.

1. Set up the oscilloscope for observing the modulated output as a trapezoidal display. To do this, set the oscilloscope for the X-Y mode of operation. Connect Y (vertical) input to v_o and the X (horizontal) input to v_{ia}. Both inputs to the oscilloscope should be set for dc input. See Figure 14-3.

2. Adjust the scaling and position of the waveform for a figure that nearly fills the screen. Adjust the amplitude of the audio source for 100% modulation. This is indicated by a trapezoidal display such as the one shown in Figure 14-4.

3. Adjust the amplitude of the audio source for 80% modulation. Sketch the resulting trapezoidal pattern in Figure 14-5 on the Results Sheet. Also indicate the measured values of V_{max} and V_{min} as determined from the display.

4. Adjust the amplitude of the audio source for 50% modulation. Sketch the resulting trapezoidal pattern in Figure 14-6 on the Results Sheet. Also indicate the measured values of V_{max} and V_{min}.

5. Increase the amplitude of the audio source to the point of 100% modulation, then continue increasing the amplitude to about 110% modulation. Sketch the resulting trapezoidal pattern in Figure 14-7 on the Results Sheet.

64

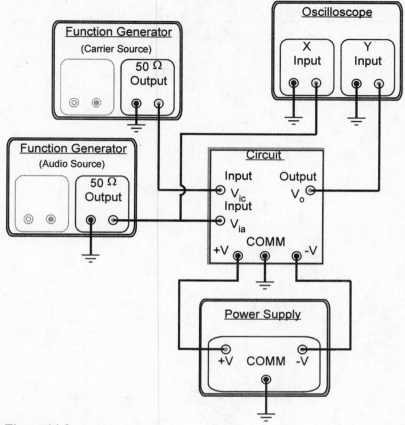

Figure 14-3

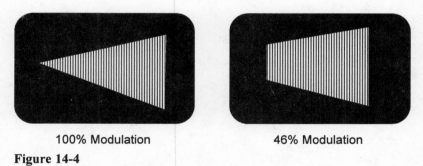

100% Modulation

46% Modulation

Figure 14-4

Experimental Notes and Calculations

Results Sheet

Project 14

Part 1   Time-Domain Modulation Display

Step 3

v_{ic} at 100% modulation = _____ V

Questions

1. What is the equation for determining percent modulation as a function of V_{max} and V_{min}?

2. Is input v_{ic} more closely associated with V_{max} or with V_{min} of Question 1? Is input v_{ia} associated with V_{max} or with V_{min}?

Part 2   Trapezoidal Modulation Display

Step 3

Measured V_{max} = _____

Measured V_{min} = _____

Step 4

Measured V_{max} = _____

Measured V_{min} = _____

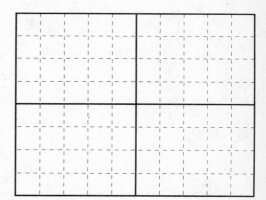

Figure 14-5

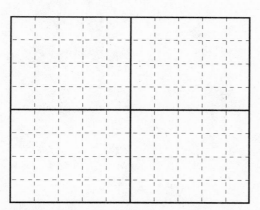

Figure 14-6

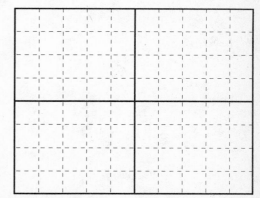

Figure 14-7

Questions

1. Which terminal on the 3080 OTA device is the inverting input? Non-inverting input? Bias?

2. How is overmodulation shown on a trapezoidal display?

Critical Thinking for Project 14

1. Explain why it is better to calculate the modulation index from observed values of V_{max} and V_{min} as viewed on the oscilloscope display than from values of inputs v_{ic} and v_{ia}.

2. Explain why integrated-circuit OTAs are not used for carrier frequencies above 500 kHz.

Project **15**

AM Spectral Analysis

An Extended Project

This project simulates a test setup that is commonly used for determining the actual frequency content of AM signals. It uses an instrument called a spectrum analyzer. This instrument provides an oscilloscope display that shows the frequency and amplitude of all signals contained within a wide band of frequencies. In this project you will:

- Adjust the frequencies and modulation index for an AM transmitter.
- Use a spectrum analyzer to determine the frequencies present in an AM signal.
- Directly observe how the frequency and amplitudes of the carrier and audio signals at the transmitter affect the content of the broadcast signal.

Preparation

Read Frenzel, *Principles of Electronic Communication Systems*, Section 3-3.

Setup Procedure

1. Select Extended Projects from the Project menu.

2. Select Project 15 AM Spectral Analysis from the list of Extended Projects.

Lab Procedure

In previous AM projects, you used an rf generator and function generator to produce the two signals required for amplitude modulation. The rf generator produced the carrier frequency and the function generator provided the audio frequency. In this project, these two instruments are combined onto a complete AM generator. With the AM generator, you can adjust the frequency of the carrier and modulation signals. Use the carrier frequency adjustment to determine the carrier frequency, and use the modulation frequency adjustment to set the modulation frequency. The modulation adjustment lets you set the modulation index between 0.0 and 1.2.

1. Set the carrier frequency to 5.0 MHz and the modulation frequency to 1000.0 kHz. Adjust the modulation to 1.0. You are now modulating a 5 MHz carrier with a 1 MHz signal at 100% modulation. Use the graph in Figure 15-1 on the Results Sheet to sketch the waveform you see on the spectrum analyzer display. Identify and label the carrier, upper sideband, and lower sideband frequencies.

2. Assume that the display of the spectrum analyzer is calibrated at 5 MHz/div on the horizontal axis, and 50 V/div on the vertical axis. Record the frequencies and amplitudes of the peaks found on the display.

3. Complete the data in Table 15-1 by setting up the indicated frequencies and modulation indexes, then recording the spectral response.

   Click the exit button when you are ready to leave this project.

Experimental Notes and Calculations

Results Sheet

Project 15

AM Spectrum

Step 2

Lower sideband:

frequency = _____, amplitude = _____

Carrier:

frequency = _____, amplitude = _____

Upper sideband:

frequency = _____, amplitude = _____

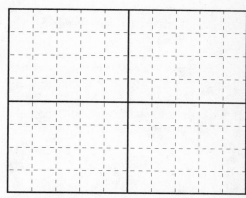

Figure 15-1

Carrier Freq (MHz)	Modulation Freq (kHz)	Modulation Index	LSB Freq	LSB Ampl	Carrier Freq	Carrier Ampl	USB Freq	USB Ampl
2.0	1000.0	1.0						
8.0	1000.0	1.0						
2.0	500.0	1.0						
5.0	500.0	1.0						
8.0	500.0	1.0						
2.0	1000.0	0.5						
5.0	1000.0	0.5						
8.0	1000.0	0.5						
5.0	1000.0	0.0						
5.0	1000.0	1.2						

Table 15-1

Questions

1. How well does the frequency spectrum of Figure 15-1 line up with the theory of AM sidebands? Explain your answer.

2. How do changes in the carrier frequency affect the spectral display?

3. How do changes in the modulation frequency affect the spectral display?

Critical Thinking for Project 15

1. Describe how changes in the modulation index affect the spectral display for an AM broadcast signal.

2. Describe the appearance of the spectral display if the modulation signal is entirely removed.

3. Describe the appearance of the spectral display if the carrier frequency is not being generated. Explain your answer.

Project **16**

Diode AM Detector

A Prep Project

In this project you will perform computer simulations of a diode AM detector circuit, noting especially the effects of the output filter capacitor. You will:

- Compare the AM signal before and after detection takes place.
- Sketch waveforms of the signals involved in the AM detection process.
- Observe the purpose of the output filter capacitor.

Preparation

Read Frenzel, *Principles of Electronic Communication Systems*, Section 4-3.

Setup Procedure

1. Select Prep Projects from the Project menu.

2. Select Project 16 Diode AM Detector from the list of Prep Projects.

Lab Procedure

In this project you will use the function generator and rf generator as signal sources for AM modulation and demodulation. The rf generator provides the carrier frequency, and the function generator provides the modulation signal. The schematic diagram shows that you are using a diode modulator and demodulator.

The schematic diagram is an interactive diagram. When you move the mouse cursor over one of the four test points, you will see the corresponding waveform on the oscilloscope display.

You can switch the order of the two diagrams by clicking the one you want in the foreground.

1. Set the amplitude of the rf generator to 6.0 V, and the amplitude of the function generator to 4.0 V.

2. Click the schematic diagram to make sure no part of it is hidden by the block diagram. Move the mouse cursor to TP 1, and sketch the oscilloscope waveform in Figure 16-1 on the Results Sheet. Move the mouse cursor to TP 2, and sketch the waveform on Figure 16-2.

3. Move the mouse cursor to TP 3 and TP 4, and sketch the corresponding waveforms on Figure 16-3.

4. Leave the amplitude of the rf generator to 6.0 V, but adjust the amplitude of the function generator to 8.0 V. Sketch in Figure 16-4 the waveforms you find at TP 3 and TP 4.

5. Leave the amplitude of the rf generator to 6.0 V, but adjust the amplitude of the function generator to 18.0 V. Sketch on Figure 16-5 the waveforms you find at TP 3 and TP 4.

Click the exit button when you are ready to leave this project.

Experimental Notes and Calculations

Results Sheet

Project 16

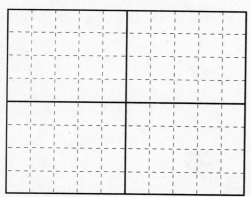

Figure 16-1

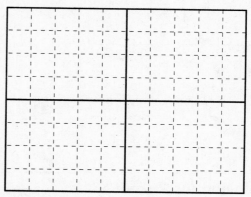

Figure 16-2

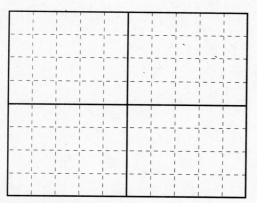

Figure 16-3

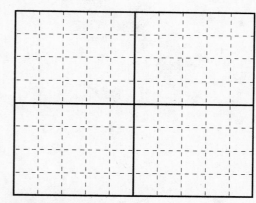

Figure 16-4

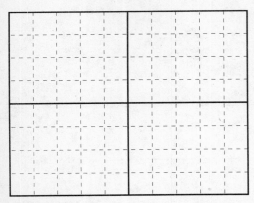

Figure 16-5

Questions

1. What is the V/div scale of the oscilloscope for this project? How did you determine that value?

2. Which, if any, of the figures that you've sketched represent an overmodulated waveform?

Critical Thinking for Project 16

1. Describe the effect that output capacitor C_2 has on the demodulated waveform. Describe how the output waveform would look if C_2 were open.

2. Describe the effect that overmodulation has on the shape of the waveform appearing at the output of the demodulator.

Project **17**

Diode AM Detector

A Hands-On Project

In this project you will study a simple AM diode detector circuit known as an *envelope detector*. You will:

- Construct both an AM modulator and demodulator.
- Compare the signal before and after detection takes place.
- Sketch waveforms of the signals involved in the AM detection process.

Components and Supplies

2   Resistor, 2.2 kΩ
1   Resistor 27 kΩ
1   Resistor, 100 kΩ
1   Capacitor, 1 μF
1   Capacitor, 10 nF
2   Capacitor, 100 nF
1   Inductor, 1 mH
1   Germanium diode, such as 1N34
1   NPN transistor, 2N2904 or equivalent

Equipment

1   Dual-trace oscilloscope
2   Function generator
1   Frequency counter (optional)

For both parts of this project, use one function generator as the audio source and the other as the carrier source.

Preparation

Read Frenzel, *Principles of Electronic Communication Systems*, Section 4-3.

Complete the work for Prep Project 16.

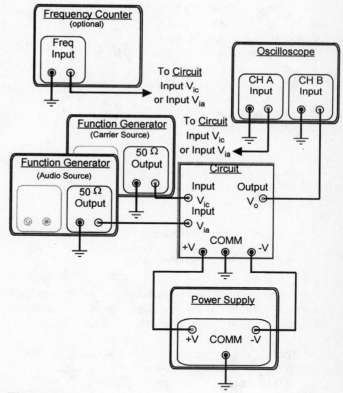

Figure 17-1

Lab Procedure

The circuit for this project is made up of the AM modulator circuit from Project 12, followed by a diode detector circuit.

1. Construct the circuit in Figure 17-2. Connect the carrier source to the v_{ic} input, and adjust it for a 1 V_{p-p} sinusoidal waveform at 500 kHz. Connect the oscilloscope to the anode of D_1, and adjust the frequency of the carrier source for a peak voltage response.

2. Connect the audio source to the v_{ia} input, and adjust it for a 400 Hz sinusoidal signal. Set the amplitude of the audio source for 100% modulation as seen at the anode of D_1. Sketch at least two cycles of the waveform on the Before axis of the graph in Figure 17-3.

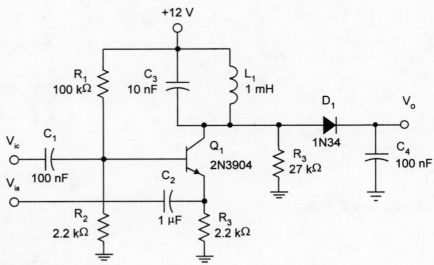

Figure 17-2

Important: The input to the oscilloscope should be set for dc input for all readings taken from output v_o.

3. Connect the oscilloscope to output v_o. Sketch at least two cycles of the output waveform on the After axis in Figure 17-3.

4. Adjust the amplitude of the audio source for 80% modulation as determined at the modulator output at v_o. Sketch at least two cycles of this waveform in Figure 17-4.

5. Adjust the amplitude of the audio source at v_{ia} for 50% modulation as determined at the modulator output at v_o. Sketch at least two of the output waveforms in Figure 17-5.

6. Set the audio source for a triangular waveform at 400 Hz. Adjust for 100% modulation as determined at the modulator output at v_o. Sketch at least two cycles of the waveform in Figure 17-6.

7. Set the audio source for a rectangular waveform at 400 Hz and 100% modulation as determined at the modulator output at v_o. Sketch at least two cycles of the waveform in Figure 17-7.

8. Set the audio source for the sinusoidal output and adjust the amplitude to overmodulate the signal. Set the modulation to a level you estimate to be about 110%. Draw at least two cycles of the output waveform in Figure 17-8.

Name ______________________________

Results Sheet

Project 17

Before										
After										

Figure 17-3

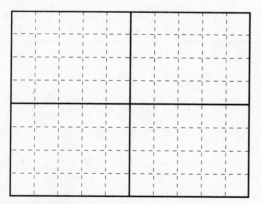

Figure 17-4

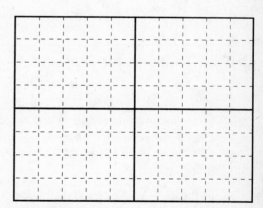

Figure 17-5

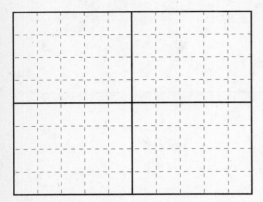

Figure 17-6

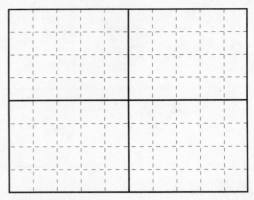

Figure 17-7

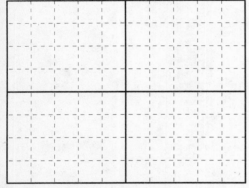

Figure 17-8

Questions

1. Which components make up the demodulator portion of this circuit?

2. How does the percent of modulation affect the amplitude of the output from an AM detector?

Critical Thinking for Project 17

1. Describe how overmodulation affects the appearance (quality) of the signal at the output of an AM demodulator circuit. Explain why overmodulation at a transmitter is undesirable.

2. Name the diode characteristic that makes it useful as an AM detector: rectification, nonlinearity, or both.

3. Explain how reversing the polarity of the detector diode would affect the operation of the detector portion of the circuit.

Project **18**

SSB Modulator

A Prep Project

This project simulates the operation of all four types of AM sideband modulation: double sideband with carrier, double sideband with no carrier, upper single sideband, and lower single sideband. You will:

- Determine the vertical and horizontal scaling of the spectrum analyzer display.
- Adjust the carrier and audio modulating frequencies for all four types of amplitude modulation.
- Read the sideband frequencies and amplitudes from a spectrum analyzer display.

Preparation

Read Frenzel, *Principles of Electronic Communication Systems*, Section 4-5.

Setup Procedure

1. Select Prep Projects from the Project menu.

2. Select Project 18 SSB Modulator from the list of Prep Projects.

Lab Procedure

The function generator provides the audio signal and the rf generator provides the carrier signal. The spectrum analyzer display provides a convenient and reliable way to determine the frequency and amplitude of the carrier and sidebands.

Part 1 Double Sideband with Carrier

Check the project bar across the top of the screen to confirm you are working with Part 1 of this project.

Note: The purpose of Step 1 is to determine the V/div vertical scaling of the spectrum analyzer display.

1. Make sure the frequency and amplitude adjustments for the function generator (audio source) are both set to zero. Set the frequency output of the rf generator to 20.0 MHz. Adjust the amplitude output of the rf generator between its two extremes, and note the carrier frequency amplitude as it appears on the spectrum analyzer display. Determine the amplitude scaling of the display by setting the rf generator amplitude for exactly one division. Record this amplitude on the Results Sheet.

Note: The purpose of Step 2 is to determine the Hz/div horizontal scaling of the spectrum analyzer display.

2. Set the output of the function generator to 500 Hz at 8.0 V. Note the upper and lower sidebands, each separated from the carrier frequency by 500 Hz. This separation means that each horizontal division on the display represents 500 Hz.

3. Set up these conditions:

   rf generator frequency = 20.0 MHz
   rf generator amplitude = 10.0 V
   function generator frequency = 500 Hz
   function generator amplitude = 10.0 V

Sketch the resulting display on the grid shown in Figure 18-1 on the Results Sheet. Also determine the values requested in that part of the Results Sheet.

4. Set up the following conditions:

       rf generator frequency = 20.0 MHz
       rf generator amplitude = 10.0 V
       function generator frequency = 750 Hz
       function generator amplitude = 5.0 V

Sketch the resulting display on Figure 18-2. Also supply the values requested in that part of the Results Sheet.

Click the browse button to go to Part 2 of this project.

Part 2   Suppressed-Carrier Double Sideband

Check the screen's project bar to confirm you are working with Part 2 of this project.

1. Set up these conditions:

       rf generator frequency = 20.0 MHz
       rf generator amplitude = 10.0 V
       function generator frequency = 500 Hz
       function generator amplitude = 10.0 V

Sketch the resulting display on Figure 18-3, and determine the values requested in that part of the Results Sheet.

2. Set up the following conditions:

       rf generator frequency = 20.0 MHz
       rf generator amplitude = 10.0 V
       function generator frequency = 750 Hz
       function generator amplitude = 5.0 V

Sketch the resulting display as Figure 18-4, and provide the values requested in that part of the Results Sheet.

Click the browse button to go to Part 3 of this project.

Part 3   Suppressed-Carrier Lower Sideband

Check the screen's project bar to confirm you are working with Part 3 of this project.

1. Set up the conditions specified earlier in Step 1 of Part 2. Record your results on Figure 18-5 of the Results Sheet.

2. Set up the conditions specified in Step 2 of Part 2. Record your results on Figure 18-6 of the Results Sheet.

Click the browse button to go to Part 4.

Part 4   Suppressed-Carrier Upper Sideband

Check the screen's project bar to confirm you are working with Part 4 of this project.

1. Set up the conditions specified earlier in Step 1 of Part 2. Record your results on Figure 18-7 of the Results Sheet.

2. Set up the conditions specified in Step 2 of Part 2. Record your results on Figure 18-8 of the Results Sheet.

Click the exit button when you are ready to leave this project.

Results Sheet

Project 18

Part 1   Double Sideband with Carrier

Step 1

    Vertical scale = _____ V/div

Step 2

    Horizontal scale = _____ Hz/div

Step 3

    Carrier amplitude = _____   Carrier frequency = _____

    USB amplitude = _____   USB frequency = _____

    LSB amplitude = _____   LSB frequency = _____

    Modulation index = _____

Step 4

    Carrier amplitude = _____   Carrier frequency = _____

    USB amplitude = _____   USB frequency = _____

    LSB amplitude = _____   LSB frequency = _____

    Modulation index = _____

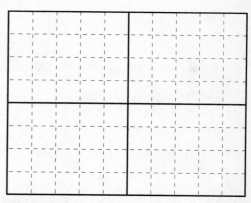

Figure 18-1

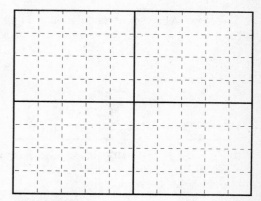

Figure 18-2

Questions

1. What portion of the FM signal are you viewing in Step 1?

2. What is the percent of modulation in Step 3?

3. What is the percent of modulation in Step 4?

Part 2   Suppressed-Carrier Double Sideband

Step 1

    USB amplitude = _____   USB frequency = _____

    LSB amplitude = _____   LSB frequency = _____

    Modulation index = _____

Step 2

    USB amplitude = _____   USB frequency = _____

    LSB amplitude = _____   LSB frequency = _____

    Modulation index = _____

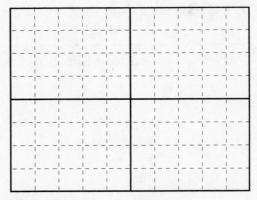

Figure 18-3

Questions

1. What is the main difference between the displays of Part 1 and Part 2?

2. What is the value of the carrier amplitude in Figure 18-3? in Figure 18-4?

3. What is the percent of modulation in Figure 18-4?

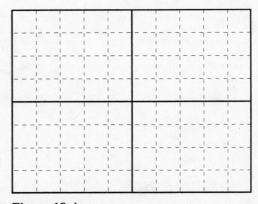

Figure 18-4

Part 3   Suppressed-Carrier Lower Sideband

Step 1

    LSB amplitude = _____   LSB frequency = _____

    Modulation index = _____

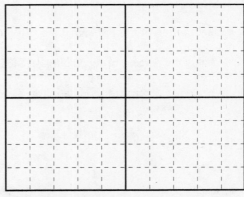

Figure 18-5

Step 2

    LSB amplitude = _____   LSB frequency = _____

    Modulation index = _____

Questions

1. What is the main difference between the displays of Part 2 and Part 3?

2. What is the value of the carrier amplitude in Figure 18-5?

3. What is the value of the USB amplitude in Figure 18-6?

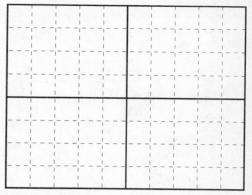

Figure 18-6

Part 4   Suppressed-Carrier Upper Sideband

Step 1

    USB amplitude = _____   USB frequency = _____

    Modulation index = _____

Step 2

    USB amplitude = _____   USB frequency = _____

    Modulation index = _____

Questions

1. What is the main difference between the displays of Part 3 and Part 4?

2. What are the similarities and the differences between the frequency and amplitude of the peaks in Figures 18-5 and 18-7?

3. What is the value of the LSB amplitude in Figure 18-8?

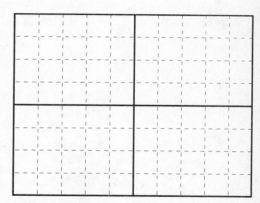

Figure 18-7

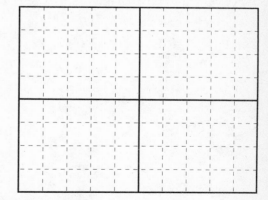

Figure 18-8

Critical Thinking for Project 18

1. Explain why it is essential to tune an SSB receiver to the same sideband that is being transmitted.

2. Name the type of filtering used in this project, and describe an alternate method.

3. Describe the primary advantage of using SSB transmission in the first place.

Project **19**

SSB Demodulator

A Hands-On Project

This project demonstrates how SSB transmissions can be fully demolulated by injecting the carrier frequency at the receiver. You will use a simple short-wave receiver and rf generator to:

- Locate SSB transmissions on the xx meter band of a conventional AM short-wave receiver.
- Note the peculiar audio quality of SSB transmission.
- Inject a carrier frequency to produce a satisfactory level of reception.

Preparation

Read Frenzel, *Principles of Electronic Communication Systems*, Section 4-5.

Complete the work for Prep Project 18.

Equipment

1   Radio receiver capable of receiving the 40- and 10-meter short-wave bands.*

1   rf generator

    * The radio receiver required for this project is commonly included as a feature on consumer tape/radio appliances, or "boom boxes."

Lab Procedure

1. Tune the receiver to the 40-meter band (7-7.3 MHz) or 10-meter (28-30 MHz) amateur band. Locate an SSB broadcast. On a conventional AM receiver, such broadcasts make a transmitted voice signal sound like someone talking in an empty barrel with a mouth full of marbles.

2. Locate the rf generator directly beside the receiver. Tune the generator through the range of frequencies for the amateur band you have selected on the receiver. Listen for the "beat" frequency (a squealing sound) as the frequency of the generator matches the frequency of the transmission you are receiving.

3. Adjust the generator to obtain the lowest pitch. You should also note that the SSB voice transmission is now more intelligible. Record the generator frequency on the Results Sheet.

4. Select another transmission on the 40- or 10-meter band and inject the carrier frequency again. Record the generator frequency for this second transmission.

Experimental Notes and Calculations

Results Sheet

Project 19

Step 3

    Generator frequency = _____

Step 4

    Generator frequency = _____

Questions

1. Assuming the modulating signal is limited to 10 kHz, what is the maximum bandwidth of an SSB transmission?

2. If the modulating signal in Question 1 is transmitted as a DSB-SC signal, what is the maximum bandwidth?

3. Assuming the modulating signal for Step 2 is 0 to 10 kHz, and that the signal is transmitted as USB-SC, what are the actual minimum and maximum frequencies?

Critical Thinking for Project 19

1. State the primary advantage and disadvantage of SSB transmission and reception.

2. AM receivers that are designed for receiving SSB transmission include a VFO (variable frequency oscillator) control. Explain the purpose of this control and relate it to the work you performed in this project.

Experimental Notes and Calculations

Project 20

Frequency Modulation

A Prep Project

In this project you will observe the operation of a circuit that changes dc and ac voltage levels into a corresponding frequency. This is the basis of frequency modulation. You will:

- Observe how the dc input level to a voltage-to-frequency converter affects the output frequency.
- Plot a curve showing output frequency as a function of dc input voltage.
- Observe the operation of a frequency modulator while varying the audio voltage level applied to it.
- Sketch an FM carrier waveform showing the effects of audio modulation.

Preparation

Read Frenzel, *Principles of Electronic Communication Systems*, Section 6-1.

Setup Procedure

1. Select Demo Projects from the Project menu.

2. Select Project 20 Frequency Modulation from the list of Prep Projects.

Lab Procedure

Part 1   DC Input

Check the project bar across the top of the screen to confirm you are working with Part 1 of this project.

This part of the project uses a dc voltage source to control the frequency of a voltage-to-frequency converter. The oscilloscope display indicates the voltage and frequency levels at the output of the circuit. A digital read-out indicates both the peak-to-peak voltage (v_o) as well as the frequency (f_o).

1. Adjust the dc voltage source for an output of -10 V (extreme left-hand setting). Then adjust it for an output of +10 V (extreme right-hand setting). Note the responses on the oscilloscope display. You do not need to record the responses at this time, however.

2. Set the dc voltage source to each of the voltage levels specified in Table 20-1 on the Results Sheet. Record the corresponding frequencies and voltages that are indicated on the oscilloscope display.

3. From the data of Table 20-1, determine the minimum frequency, maximum frequency, frequency when $v_i = 0$, and the maximum frequency deviation. Record your answers on the Results Sheet.

4. From the data of Step 3, calculate the frequency/volt ratio for this circuit. Record your answer on the Results Sheet.

   Click the browse button to go to Part 2 of this project.

Part 2 Audio Input

Check the screen's project bar to confirm you are working with Part 2 of this project.

In this part of the project, a function generator applies a sinusoidal ac waveform to the input of the modulation. You are able to vary the amplitude and frequency of that waveform and to note the response on the oscilloscope screen. As indicated on the block diagram for Part 2, the upper trace on the oscilloscope shows the ac input to the modulator. The lower trace shows the frequency-modulated output.

1. Set the amplitude of the function generator to 5 V. Adjust the Frequency to produce two full cycles on the upper trace of the oscilloscope display (about 12 kHz).

2. Set the amplitude to its minimum level (1 V). Then step the input voltage upward one volt at a time by clicking the right arrow on the amplitude control. Notice how the level of the input signal increases. As the input signal level approaches 10 V, notice how the time between peaks on the output signal varies with the input sine wave.

3. Make certain the frequency of the function generator is still set at about 12 kHz. Set the amplitude to the maximum of 10 V. Sketch the upper and lower traces of the oscilloscope display, using the graph in Figure 20-1 on the Results Sheet.

4. Set the frequency of the function generator to its maximum of 40 kHz. Set the amplitude to the minimum of 1 V.

5. Step the input voltage upward one volt at a time by clicking the right arrow on the amplitude control. Notice how the time between peaks on the output signal varies with the input sine wave.

Click the exit button when you are ready to leave this project.

Results Sheet

Project 20

Part 1   DC Input

Step 3

     minimum frequency = _____

     maximum frequency = _____

     frequency at 0-V input = _____

     maximum frequency deviation = _____

Step 4

     frequency/volt = _____

Questions

1. How does the amount of dc input voltage affect the peak-to-peak output voltage of this circuit?

2. Suppose this circuit is altered so that the output is 1500 kHz when $v_i = 0$. Assuming the frequency/volt figure remains unchanged, what will be the minimum and maximum frequencies for this circuit? the maximum frequency deviation?

dc Voltage In	Frequency Out	Voltage Out
-10		
-8		
-6		
-4		
-2		
0		
2		
4		
6		
8		
10		

Table 20-1

Part 2   Audio Input

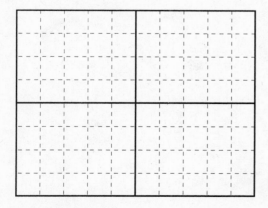

Figure 20-1

Questions

1. How does changing the amplitude of the modulating signal affect the amplitude of the output signal?

2. In what way does changing the frequency of the modulating signal affect the amplitude of the output signal?

Critical Thinking for Project 20

1. In the circuit of Part 1, the output frequency increases as the input dc level rises in a positive direction. Suppose, however, you want a circuit that shows an increasing output frequency as the dc input level goes more negative. Describe where you would place an inverting dc amplifier in the block diagram in order to make the circuit operate that way.

2. The oscilloscope display for Part 2 of this project shows the FM waveform when the modulating signal is a sine wave. Sketch the same set of waveforms where the modulating signal is a rectangular waveform.

Project **21**

Frequency Modulation

A Hands-On Project

This project uses a voltage-controlled oscillator (VCO) as a frequency modulator. You will:

- Construct the circuit.
- Vary the dc input voltage level and record the corresponding output frequency.
- Plot a graph showing output frequency as a function of input voltage level.

Preparation

Read Frenzel, *Principles of Electronic Communication Systems*, Section 6-1.

Complete the work for Prep Project 20.

Components and Supplies

2	Resistor, 1 kΩ
1	Resistor, 2.2 kΩ
1	Potentiometer, 10 kΩ
1	Capacitor, 1 nF
1	IC, NC566 Voltage-controlled oscillator

Equipment

1	dc power supply
1	Function generator
1	Dual-trace oscilloscope
1	Digital voltmeter (optional)
1	Frequency counter (optional)

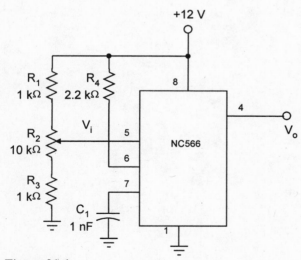

Figure 21-1

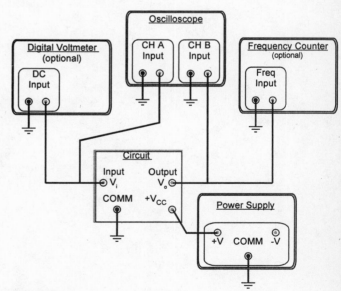

Figure 21-2

Lab Procedure

1. Construct the voltage-to-frequency converter circuit in Figure 21-1. Connect the oscilloscope to monitor the dc input at v_i (pin 5) and the signal output at v_o (pin 4). Connect other equipment as shown in Figure 21-2.

2. Adjust the potentiometer for each of the dc voltage levels shown in Table 21-1 on the Results Sheet. Record the corresponding v_i voltage and v_o frequency in the table.

3. Plot the data of Table 21-1, showing the output frequency as a function of input voltage level. Use the graph in Figure 21-3 on the Results Sheet for this plot.

Results Sheet

Project 21

dc Voltage In	Frequency Out
2	
4	
6	
8	
10	

Table 21-1

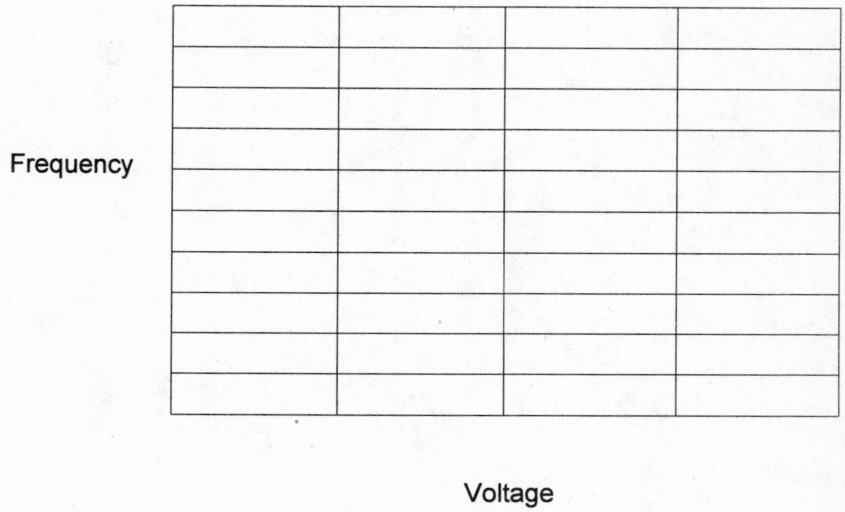

Frequency

Voltage

Figure 21-3

Questions

1. From the data in Table 21-1, what is the sensitivity of this modulator in terms of kHz/V?

2. Does the output frequency increase or decrease as the modulating voltage goes more positive?

Critical Thinking for Project 21

1. Name the two passive components in Figure 21-1 that determine the center operating frequency of the oscillator.

2. Describe the effect that increasing the value of C_1 from 1 nF to 10 nF will have upon the operation of the circuit.

3. Explain how you would use the 566 VCO to make a sweep-frequency function generator.

Experimental Notes and Calculations

Project 22

Frequency Demodulation

A Prep Project

This project is a computer simulation of tests on a frequency-to-voltage converter. You will:

- Apply various frequency levels to the input of the circuit, and observe the resulting output voltage levels.
- Demonstrate the effect that the input amplitude has on the output voltage levels.
- Plot graphs based on the data gathered from the circuit.

Preparation

Read Frenzel, *Principles of Electronic Communication Systems*, Section 6-3.

Setup Procedure

1. Select Prep Projects from the Project menu.

2. Select Project 22 Frequency Demodulation from the list of Prep Projects.

Lab Procedure

This project uses a function generator to supply various frequencies and voltage levels to a frequency-to-voltage converter circuit. A dc voltmeter monitors the output of the circuit.

1. Set the amplitude of the function generator to 5 V. Adjust the frequency of the function generator to each of the levels shown in Table 22-1 on the Results Sheet. Record the corresponding output voltage levels.

2. Plot a graph showing how the output voltage changes with the input frequency. Use the graph shown in Figure 22-1.

3. With the amplitude of the function generator still set at 5 V, adjust the frequency to obtain an output of 0.0 V. Record this frequency on the Results Sheet.

4. Set the amplitude of the function generator to each of the levels shown in Table 22-2. Record the corresponding output voltage.

5. Plot a graph that shows how the output voltage changes with the input voltage. Use the space provided in Figure 22-2.

Click the exit button when you are ready to leave this project.

Experimental Notes and Calculations

Results Sheet

Project 22

Frequency In (MHz)	Voltage Out (V)
400	
410	
420	
430	
440	
450	
460	
470	
480	
490	
500	

Table 22-1

Voltage

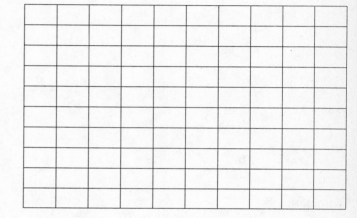

Frequency

Figure 22-1

Step 3

Frequency input for 0-V output = _____

Voltage In (V)	Voltage Out (V)
1	
2	
3	
4	
5	
6	
7	
8	
9	
10	

Table 22-2

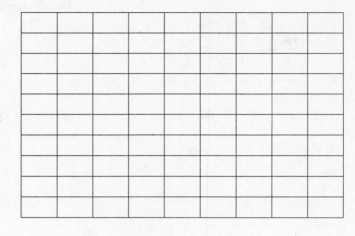

Voltage Out

Voltage In

Figure 22-2

Questions

1. From the data in Figure 22-1, what is the average V/MHz conversion?

2. From the results in Figure 22-2, what effect does a change in input voltage have upon the output voltage?

Critical Thinking for Project 22

1. Explain how the circuit for this project can be used as an FM detector.

2. Suppose the frequency applied to this circuit is increasing at a steady rate. Describe how the output voltage responds.

3. Suppose the frequency applied to this circuit drops to 0 Hz. Describe the effect you would most likely notice at the dc output of the circuit.

Project **23**

Frequency Demodulation

A Hands-On Project

This project uses a phase-locked-loop (PLL) device to demodulate an FM signal generated by the simple modulator used in Project 22. You will:

- Construct the PLL demodulator circuit.
- Determine the VCO frequency of the PLL.
- Plot dc output as a function of input frequency.
- Observe the FM demodulation of a sweep-frequency signal.

Components and Supplies

1	Resistor, 4.7 kΩ
1	Resistor, 47 kΩ
1	Capacitor, 1 nF
1	Capacitor, 10 nF
1	Capacitor, 100 nF
1	IC, LM565 phase-locked loop

Preparation

Read Frenzel, *Principles of Electronic Communication Systems*, Section 6-1.

Complete the work for Prep Project 22.

Equipment

1	Dual-voltage dc power supply
1	Function generator with a sweep-frequency mode
1	Dual-trace oscilloscope

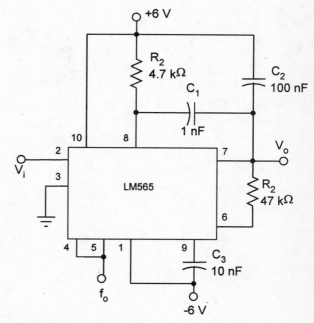

Figure 23-1

Lab Procedure

1. Construct the FM demodulator circuit shown in Figure 23-1. **Do not** at this time connect the function generator to input v_i.

2. Calculate the frequency of the VCO from the component values shown in Figure 23-1. Measure the actual VCO frequency at f_o. Record both the calculated and measured frequencies on the Results Sheet.

3. Adjust the function generator for a sinusoidal output. Set the frequency to the measured value of f_o from Step 2, and set the amplitude to 2 $V_{p\text{-}p}$. Apply this signal to input v_i of the FM demodulator circuit. Connect the oscilloscope to monitor v_i on one channel and v_o on the other. Record the frequency at v_i and the dc output voltage at v_o.

4. Decrease the frequency at v_i by 10% from the frequency used in Step 3. Record this frequency and the dc output voltage at v_o.

5. Increase the frequency at v_i by 10% from the frequency used in Step 3. Record this frequency and the dc output voltage at v_o.

6. Plot the data (v_o as a function of f_i) from Steps 4 and 5 on the graph provided in Figure 23-2.

7. Switch the operating mode of the function generator to produce a sweep-frequency output. Set the generator's sweep rate to the highest available frequency and the sweep range to maximum. Observe the output at v_o and trigger the oscilloscope from the same signal. Make any necessary slight adjustments in the function generator's sweep range in order to produce the cleanest triangular waveform on the oscilloscope display. This is the FM-demodulated version of the sweep-frequency signal at v_i. Sketch the waveform in Figure 23-3.

Results Sheet

Project **23**

Step 2

      Calculated f_o = _____
      Measured f_o = _____

Step 3

      f_i = _____
      v_o = _____

Step 4

      f_i = _____
      v_o = _____

Step 5

      f_i = _____
      v_o = _____

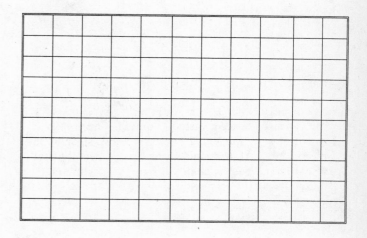

Figure 23-2

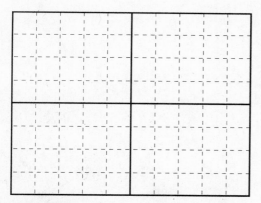

Figure 23-3

Questions

1. What is the formula for estimating the frequency of the VCO in the PLL device when there is no input signal applied?

2. Is the dc output at v_o proportional or inversely proportional to the input frequency?

Critical Thinking for Project 23

1. Describe how the circuit arrangement in Step 7 is doing the job of an FM demodulator.

2. Suppose the sweep range of the input signal in Step 7 exceeds the lower lock range of the PLL circuit. Describe where the resulting distortion will appear on the output waveform at v_o.

Experimental Notes and Calculations

Project **24**

FM Spectrum Analysis

An Extended Project

This project simulates a test setup for directly observing the carrier frequency and sidebands of an FM broadcast signal. You will:

- Calculate the modulation index.
- Observe the carrier and sideband frequencies.
- Verify the Bessel function table.
- Determine the bandwidth.

This computer simulation solves the Bessel functions that describe the frequency spectrum of an FM broadcast signal.

Preparation

Read Frenzel, *Principles of Electronic Communication Systems*, Section 5-3.

Setup Procedure

1. Select Extended Projects from the Project menu.

2. Select Project 24 FM Spectrum Analysis from the list of Extended Projects.

Lab Procedure

The instrument shown as an FM generator gives you complete control over the FM signal. The oscilloscope display for a spectrum analyzer shows the sideband frequencies and their amplitudes.

The two controls on the FM generator allow you to adjust the signal's frequency deviation and carrier frequency. While experimenting with these adjustments, you might discover that they interact with one another under certain circumstances. Actually, the controls are designed so that the setting for the frequency deviation can be no more than four times the setting for the carrier frequency. In other words, the instrument is fixed so that the modulation index of the FM signal cannot exceed 4.

The spectral display includes digital readouts for frequency and amplitude. To use these readouts, move the mouse pointer to the place on the screen where you want to determine frequency and amplitude. Those values automatically appear on the digital readouts.

1. Set the frequency deviation to 0.00 MHz and the carrier frequency to 125.00 MHz. Sketch the spectrum analyzer display in Figure 24-1 on the Results Sheet. Move the mouse pointer to the peak frequency on the display and record the peak amplitude and frequency. Calculate and record the modulation index for this signal.

2. Set the frequency deviation to 150.00 MHz and the carrier frequency to 150.00 MHz. Sketch the spectrum analyzer display as Figure 24-2 on the Results Sheet. Determine the amplitude and frequency of each peak, and indicate your values on the figure. Calculate and record the modulation index.

3. Set the frequency deviation to 300.00 MHz and the carrier frequency to 100.00 MHz. Sketch the spectrum analyzer display as Figure 24-3 on the Results Sheet. Determine the amplitude and frequency of each peak, and indicate your values on the figure. Calculate and record the modulation index.

4. Set the frequency deviation to 1200.00 MHz and the carrier frequency to 300.00 MHz. Sketch the spectrum analyzer display as Figure 24-4 on the Results Sheet. Determine the amplitude and frequency of each peak, and indicate your values on the figure. Calculate and record the modulation index.

Click the exit button when you are ready to leave this project.

Experimental Notes and Calculations

Results Sheet

Project 24

Step 1

    peak voltage = _____   peak frequency = _____

    modulation index = _____

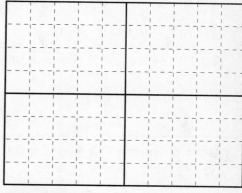

Figure 24-1

Step 2

    modulation index = _____

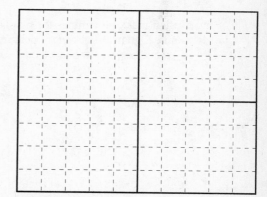

Figure 24-2

Step 3

    modulation index = _____

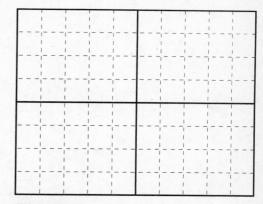

Figure 24-3

Step 4

modulation index = _____

Questions

1. According to your table of Bessel functions, how many different frequencies should be present for each of the modulation indexes used in this project?

2. How well do each of the voltage levels in Figure 24-4 compare with the voltages cited in your table of Bessel functions, for a signal of the same modulation index?

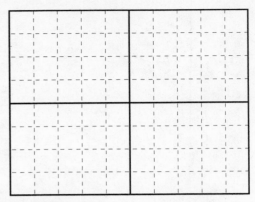

Figure 24-4

Critical Thinking for Project 24

1. Explain the meaning of FM sideband and carrier peaks that have negative values.

2. The modulation index determines the number of significant sidebands. Explain what factors determine the frequency difference between the individual sidebands.

3. Describe how you could use the simulated instruments of this project to verify the Bessel function curves shown in your textbook.

Project 25

Varactor Modulator

An Extended Project

This project uses simulated devices, circuits, and instruments to demonstrate the operation of a varactor diode. You will:

- Gather data and plot a curve showing how the amount of reverse voltage applied to a varactor diode affects its capacitance.
- Gather data and plot response curve for a voltage-to-frequency converter circuit that uses a varactor diode.
- Measure the amount of phase shift caused by changing the amount of reverse bias on a varactor diode operating in a phase modulation circuit.

Preparation

Read Frenzel, *Principles of Electronic Communication Systems*, Section 6-1.

Setup Procedure

1. Select Extended Projects from the Project menu.

2. Select Project 25 Varactor Modulator from the list of Extended Projects.

Lab Procedure

Part 1   DC Capacitance Control

Check the project bar across the top of the screen to confirm you are working with Part 1 of this project.

In this part of the project, you apply a varying amount of dc voltage to a varactor diode and directly read the capacitance. The voltage is supplied by an adjustable dc voltage source. The corresponding capacitance of the varactor diode is monitored by a simulated capacitance meter.

1. Set the dc voltage source to the values listed in Table 25-1 on the Results Sheet. Record the corresponding capacitance values for the varactor diode.

2. Plot the data of Table 25-1 on the graph provided in Figure 25-1.

   Click the browse button to go to Part 2.

Part 2   VCO Operation

Check the screen's project bar to confirm you are working with Part 2 of this project.

In this part of the project, the varactor diode is connected as the capacitive part of an *LC* oscillator. A dc voltage applied to the varactor diode causes its capacitance to change and, therefore, causes the frequency of the oscillator to change as well.

1. Set the dc voltage source to the values listed in Table 25-2 on the Results Sheet. Record the corresponding capacitance values for the varactor diode.

2. Plot the data of Table 25-2 on the graph provided as Figure 25-2.

   Click the browse button to go to Part 3.

Part 3   PM Operation

Check the screen's project bar to confirm you are working with Part 3 of this project.

In this part of the project, you are applying a dc voltage to a phase modulation circuit. You can view a schematic diagram of the circuit by clicking the schematic button, located on the right-hand side of the button bar. Also click the schematic button when you want to remove the schematic from the work area.

Refer to the oscilloscope display. The upper waveform shows the applied voltage of the generator. This is the reference waveform. The lower waveform shows the output waveform from the phase-shift circuit--the PM signal. This waveform is shifted with respect to the upper waveform. Moving the mouse pointer across the oscilloscope screen

causes a digital display to indicate the phase angle of the reference waveform. Use this feature to determine the phase shift of the PM signal.

1. Move the control on the dc voltage source between its extreme settings. Notice how the phase of the lower waveform is shifted with respect to the upper waveform.

2. Set the dc voltage source control to the values shown in Table 25-3 on the Results Sheet. Estimate the phase angle of the shifted waveform, relative to the reference waveform. Record your results on the same table.

3. Plot a curve on Figure 25-3, showing how the dc input voltage level affects the amount of phase shift.

Click the browse button to go to Part 4.

Part 4   PM Operation with Inversion

Check the screen's project bar to confirm you are working with Part 4 of this project.

1. Move the control on the dc voltage source between its extreme settings. Notice how the phase of the lower waveform is shifted with respect to the upper waveform.

2. Set the dc voltage source control to the values shown in Table 25-4 on the Results Sheet. Estimate the phase

angle of the shifted waveform, relative to the reference waveform. Record your results on the same table.

3. Plot a curve on Figure 25-4, showing how the dc input voltage level affects the amount of phase shift.

Click the exit button when you are ready to leave this project.

Name _____

Results Sheet

Project **25**

Part 1   DC Capacitance Control

Voltage In (V)	Capacitance (pF)
-2	
-4	
-6	
-8	
-10	
-12	
-14	
-16	
-18	
-20	

Table 25-1

Capacitance

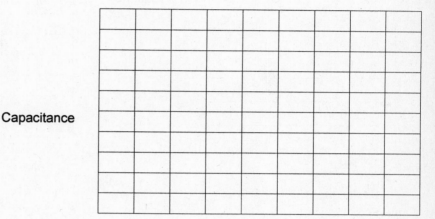

Voltage In

Figure 25-1

Questions

1. From the graph in Figure 25-1, what voltage must be applied to produce a capacitance of 200 pF?

2. According to the data of Figure 25-1, what is the pF/V rating of this varactor diode in the -2 V to -4 V range?

Part 2   VCO Operation

Voltage In (V)	Frequency Out (MHz)
-2	
-4	
-6	
-8	
-10	
-12	
-14	
-16	
-18	
-20	

Table 25-2

Frequency Out

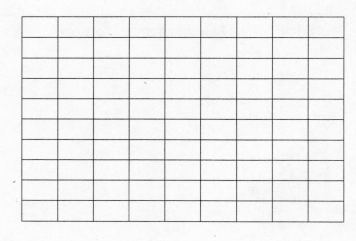

Voltage In

Figure 25-2

Questions

1. From the graph in Figure 25-2, what voltage must be applied to produce a frequency of 15.5 MHz?

2. According to the data of Figure 25-2, what is the approximate MHz/V rating of this VCO?

Part 3   PM Operation

Voltage In (V)	Phase Angle* (deg)
-2	
-4	
-6	
-8	
-10	
-12	
-14	
-16	
-18	
-20	

Table 25-3

Phase Angle (deg)

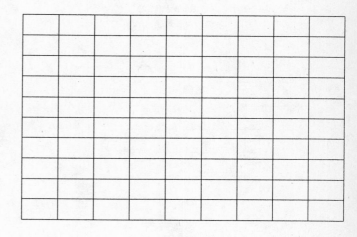

Voltage In

Figure 25-3

*Use a minus sign (–) to indicate a lagging phase angle, and a plus sign (+) to indicate a leading phase angle.

Questions

1. Does the lower waveform lead or lag the upper waveform?

2. What dc voltage is required for a phase shift of 57°?

3. What is the conversion (degree/volt) rating of this circuit in the range of -8 V to -20 V?

Part 4   PM Operation with Inversion

Voltage In (V)	Phase Angle* (deg)
-2	
-4	
-6	
-8	
-10	
-12	
-14	
-16	
-18	
-20	

Table 25-4

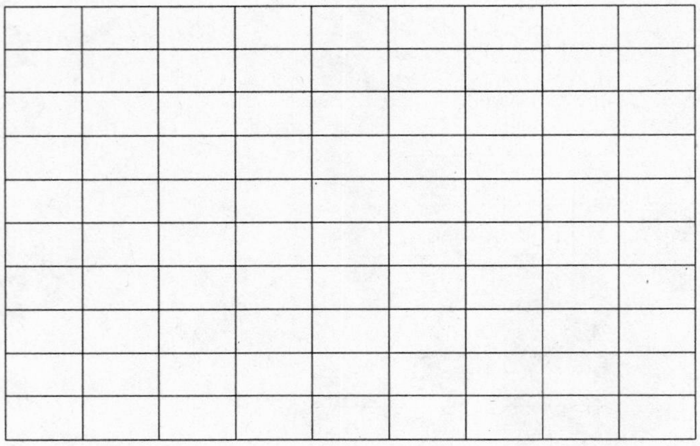

Phase Angle (deg)

Voltage In

Figure 25-4

*Use a minus sign (−) to indicate a lagging phase angle, and a plus sign (+) to indicate a leading phase angle.

Questions

1. Does the lower waveform lead or lag the upper waveform?

2. What dc voltage is required for a phase shift of 64°?

3. What is the conversion (degree/volt) rating of this circuit in the input range of -2 V to -14 V?

Critical Thinking for Project 25

1. Explain the operation of a varactor diode in terms of the width of its depletion layer.

2. Describe the essential differences in the theory of operation between ceramic filters and varactor diodes.

3. This project assumes the varactor is being operated in its linear region. Explain what this assumption means.

Project **26**

PLL Operation

A Prep Project

This project simulates the operation of a phase-locked loop (PLL). An RF generator, frequency counter, and DVM provide all the instrumentation required for studying this PLL circuit. You will:

- Determine the free-running frequency of a PLL.
- Determine the lock range of the PLL circuit.
- Determine the capture range of the PLL circuit.
- Observe how changing the input frequency affects the VCO frequency and error voltage level.

Preparation

Read Frenzel, *Principles of Electronic Communication Systems*, Section 6-3.

Setup Procedure

1. Select Prep Projects from the Project menu.

2. Select Project 26 PLL Operation from the list of Prep Projects.

Lab Procedure

When you first start this simulation, notice the black rectangle located on the circuit icon. This represents an LED (light-emitting diode) that is sensitive to the circuit's capture mode of operation. This lamp is red when the PLL is in its capture mode. For the purposes of this project, indicate OFF when the lamp is black and indicate ON when the lamp is red.

1. Set the rf generator for its minimum output frequency. Record the readings on all three instruments. Note the status of the LED on the circuit.

2. Click the right arrow button on the frequency adjustment of the generator to increase the frequency at 0.1 MHz steps. Continue this stepping operation until the lamp goes ON (turns to red). At that point, record the status of the instruments.

Note: Do not reduce the frequency setting through Steps 3 and 4. If you happen to overshoot a reading, do not attempt to decrease the frequency in order to recover it. Instead, begin the sequence again at Step 1.

3. Resume clicking the right arrow button on the frequency adjustment of the generator to increase the frequency at 0.1 MHz steps. Continue this stepping operation until the lamp goes OFF (turns to black). At that point, record the status of the instruments.

4. Resume clicking the right arrow button on the frequency adjustment of the generator, until you reach the maximum frequency output. Once more, record the status of the instruments.

5. Now begin clicking the left arrow button on the frequency adjustment of the generator to *decrease* the frequency at 0.1 MHz steps. Continue this stepping operation until the lamp goes ON. At that point, record the status of the instruments.

Note: Do not increase the frequency setting through Steps 6 and 7. If you step down beyond a point where you are supposed to stop for a set of readings, restart the operation from Steps 4 and 5.

6. Resume clicking the left arrow button on the frequency adjustment of the generator to decrease the frequency at 0.1 MHz steps, until the lamp goes OFF. At that point, record the status of the instruments.

7. Continue clicking the left arrow button on the frequency adjustment of the generator until you reach the minimum frequency output once again. Record the final status of the instruments.

Click the exit button when you are ready to leave this project.

Experimental Notes and Calculations

Name _____

Results Sheet

Project 26

Step 1

    Frequency generator = _____

    Frequency counter = _____

    Digital voltmeter = _____

    Circuit lamp status (ON or OFF) = _____

Step 2

    Frequency generator = _____

    Frequency counter = _____

    Digital voltmeter = _____

    Circuit lamp status (ON or OFF) = _____

Step 3

    Frequency generator = _____

    Frequency counter = _____

    Digital voltmeter = _____

    Circuit lamp status (ON or OFF) = _____

Step 4

    Frequency generator = _____

    Frequency counter = _____

    Digital voltmeter = _____

    Circuit lamp status (ON or OFF) = _____

Step 5

    Frequency generator = _____

    Frequency counter = _____

    Digital voltmeter = _____

    Circuit lamp status (ON or OFF) = _____

Step 6

    Frequency generator = _____

    Frequency counter = _____

    Digital voltmeter = _____

    Circuit lamp status (ON or OFF) = _____

Step 7

    Frequency generator = _____

    Frequency counter = _____

    Digital voltmeter = _____

    Circuit lamp status (ON or OFF) = _____

Questions

1. What is the free-running frequency of this PLL?

2. What are the frequencies at the lower and upper limits of the lock range?

3. What are the frequencies at the lower and upper limits of the capture range?

Critical Thinking for Project 26

1. Explain why the LED turned ON at one frequency (when you were stepping the frequency upward), but it went OFF at a different frequency (when you were stepping the frequency downward).

2. Explain why the LED turned OFF at one frequency (when you were stepping the frequency upward), but it went ON at a different frequency (when you were stepping the frequency downward).

3. Explain why an oscilloscope would be a better instrument for monitoring the error output voltage if you were using a real-time FM signal at the input of the circuit.

Project 27

PLL Operation

A Hands-On Project

The purpose of this project is to investigate the several modes of operation of a phase-locked-loop PLL circuit. You will:

- Construct the circuit.
- Determine the VCO frequency.
- Evaluate the capture/lock frequency hysteresis.

Components and Supplies

1	Resistor, 4.7 kΩ
1	Resistor, 47 kΩ
1	Capacitor, 1 nF
1	Capacitor, 10 nF
1	Capacitor, 100 nF
1	IC, LM565 PLL

Equipment

1	Dual-voltage dc power supply
1	Function generator
1	Dual-trace oscilloscope

The frequency counter is optional because the frequencies for this project can be determined (although somewhat less conveniently) with the oscilloscope.

Preparation

Read Frenzel, *Principles of Electronic Communication Systems*, Section 6-3.

Complete the work for Prep Project 26.

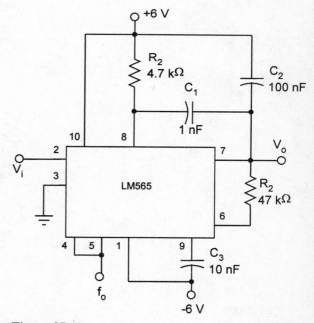

Figure 27-1

Lab Procedure

1. Construct the circuit shown in Figure 27-1. Apply power and measure the frequency of the signal at output f_o. Record the result on the Results Sheet.

2. Connect the oscilloscope to monitor the frequency at v_i on one trace and f_o on the other. Set the function generator for:

   Rectangular waveform
   2 $V_{p\text{-}p}$ amplitude
   Frequency equal to the frequency of Step 1

   Connect the function generator to input v_i. Trigger the oscilloscope from the signal at v_i, and adjust the display to view both waveforms clearly. Sketch the waveform in Figure 27-2.

3. Gradually increase the frequency of the function generator, noting how the phase changes between the two waveforms. At some point, the PLL will lose its lock on the input waveform—the signal from f_o will suddenly appear unstable on the oscilloscope. Record at v_i the frequency at which this happens.

4. Gradually decrease the frequency of the function generator, noting the point where the two waveforms lock once again. Record at v_i the frequency where this relocking occurs.

5. Continue decreasing the frequency of the function generator until lock is lost again. Record the input frequency.

6. Gradually increase the frequency of the function generator until the lock is restored. Record input frequency at this point.

Results Sheet

Project 27

Step 2

No-signal f_o = _____

Step 3

Frequency at v_i = _____

Step 4

Frequency at v_i = _____

Step 5

Frequency at v_i = _____

Step 6

Frequency at v_i = _____

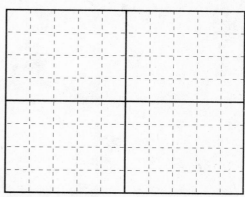

Figure 27-2

Questions

1. According to the data from Steps 3 through 6, what are the upper and lower lock frequencies of this circuit?

2. According to the data from Steps 3 through 6, what are the upper and lower capture frequencies of this circuit?

Critical Thinking for Project 27

1. Explain why it is important to keep the input frequency within the locking range of a PLL when the circuit is being used as an FM demodulator.

2. Describe the effect that a change of ± 20% in the amplitude at v_i would have upon the signal at v_o.

Experimental Notes and Calculations

Project **28**

Crystal Oscillator

A Prep Project

This project is a computer simulation of tests on a JFET Pierce crystal oscillator and a TTL crystal oscillator. You will:

- Determine the operating frequency, given the values of the circuit components.
- Determine the actual operating frequency from an oscilloscope display.
- Observe phase relationships in the TTL version.

Preparation

Read Frenzel, *Principles of Electronic Communication Systems*, Section 7-2.

Setup Procedure

1. Select Prep Projects from the Project menu.

2. Select Project 28 Crystal Oscillator from the list of Prep Projects.

Lab Procedure

This project uses interactive schematic diagrams. Moving the mouse pointer to one of the test points on the diagram simulates touching that point with an oscilloscope probe.

The waveform for that point appears on the oscilloscope display. Also on the display you will find digital readouts for the frequency and amplitude of the waveform.

Part 1   Pierce Oscillator

Check the project bar across the top of the screen to confirm you are working with Part 1 of this project.

1. Determine the voltage and frequency at all four test points. Record your findings on the Results Sheet.

2. Determine the phase differences between the signals at TP 1 and TP 2; TP 2 and TP 3; TP 3 and TP 4. Record your data on the results sheet.

Click the browse button to go to Part 2 of this project.

Part 2   TTL Crystal Oscillator

Check the screen's project bar to confirm you are working with Part 2 of this project.

1. Determine the voltage and frequency at all four test points. Record your findings on the Results Sheet.

2. Determine the phase differences between the signals at TP 1 and TP 2, TP 2 and TP 3, TP 3 and TP 4. Record your data on the Results Sheet.

Click the exit button when you are ready to leave this project.

Results Sheet

Project 28

Part 1   Pierce Oscillator

Step 1

TP 1 voltage = _____

TP 1 frequency = _____

TP 2 voltage = _____

TP 2 frequency = _____

TP 3 voltage = _____

TP 3 frequency = _____

Step 2

Phase difference between TP 1 and TP 2: _____

Phase difference between TP 2 and TP 3: _____

Questions

1. What is the phase relationship between the signals at TP 1 and TP 2?

2. What is the phase relationship between the signals at TP 2 and TP 3?

Part 2   TTL Crystal Oscillator

Step 1

TP 1 voltage = _____

TP 1 frequency = _____

TP 2 voltage = _____

TP 2 frequency = _____

TP 3 voltage = _____

TP 3 frequency = _____

TP 4 voltage = _____

TP 4 frequency = _____

Step 2

Phase difference between TP 1 and TP 2: _____

Phase difference between TP 2 and TP 3: _____

Phase difference between TP 3 and TP 4: _____

Questions

1. Which components on the schematic diagram determine the operating frequency of this circuit?

2. What is the voltage gain of the circuit between points TP 1 and TP 2? between TP 3 and TP 4?

Critical Thinking for Project 28

1. Describe the operation of a crystal oscillator in a circuit where the amplifier stage has a voltage gain of less than 1.

2. Explain how IC_1-A, IC_1-B, and XTAL in the circuit for Part 2 comprise a positive feedback loop.

3. Describe the operation of the circuit for Part 2 if the feedback loop included IC_1-C as well as IC_1-A, IC_1-B, and the crystal.

Project **29**

Crystal Oscillator

A Hands-On Project

The circuits in this project use a JFET as the active element of a Pierce oscillator, and TTL logic inverters as the active elements in a simple crystal oscillator. You will:

- Construct the circuits.
- Measure operating frequencies.
- Gather data for determining the oscillators' frequency stability.

Components and Supplies

2	Resistor, 330 Ω
1	Resistor, 100 kΩ
1	Capacitor, 10 nF
1	Capacitor, 100 nF
2	Capacitor, 100 pF
1	JFET, 2N5457
1	IC, 7404 TTL hex inverter
1	Crystal, 3.579 MHz
1	Crystal, 5.00 MHz

Preparation

Read Frenzel, *Principles of Electronic Communication Systems*, Section 7-2.

Complete the work for Prep Project 28.

Equipment

1	Variable dc power supply
1	Oscilloscope
1	Frequency counter (optional)

Lab Procedure

Part 1　Pierce Crystal Oscillator

1. Construct the Pierce oscillator circuit shown in Figure 29-1. Adjust the supply voltage for +12 V, then measure and record the frequency and signal level at v_o.

2. For each of the entries in Table 29-1, set the power supply to the given voltage level and record the frequency and peak-to-peak value of the waveform at v_o.

3. Plot the frequency data of Step 2 on the graph in Figure 29-3.

4. Replace the 3.578 MHz crystal in the circuit with a 5.00 MHz crystal. Measure and record the frequency and signal level at v_o.

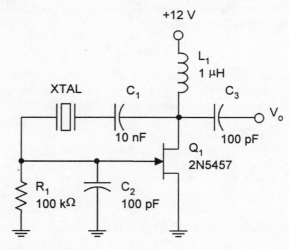

Figure 29-1

Part 2　TTL Crystal Oscillator

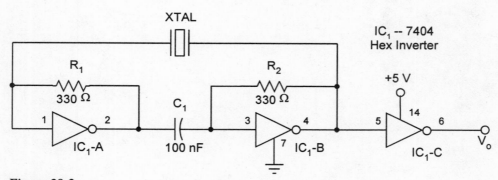

Figure 29-2

1. Construct the TTL oscillator circuit shown in Figure 29-2. Make sure you use the +5 Vdc terminals on the power supply, and apply power to the circuit. Connect the oscilloscope to output v_o, then measure and record the frequency and signal level at v_o.

2. Observe the waveforms at the outputs of IC_1-A, IC_1-B, and IC_1-C. Sketch these three waveforms in the spaces provided in Figure 29-4, making sure your drawings show the phase relationships among these three points in the circuit.

Results Sheet

Project 29

Part 1   Pierce Crystal Oscillator

Step 1

$f_o =$ _____ $v_o =$ _____ $V_{p\text{-}p}$

Step 4

$f_o =$ _____ $v_o =$ _____ $V_{p\text{-}p}$

Questions

1. Based on the data in Figure 29-4, can you say that a Pierce crystal oscillator is frequency stable with regard to changes in the power supply voltage? Explain your answer.

2. Assuming that the 5.00 MHz crystal in this project has a rated precision of ± 50 ppm, what are the minimum and maximum allowable frequencies?

dc Supply Voltage	Output Voltage ($V_{p\text{-}p}$)	Output Frequency
+2 V		
+4 V		
+6 V		
+8 V		
+10 V		
+12 V		

Table 29-1

Output Frequency

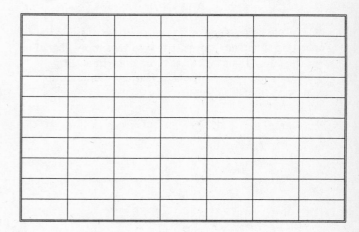

Supply Voltage

Figure 29-3

Part 2　　TTL Crystal Oscillator

Step 1

$f_o =$ _____

$v_o =$ _____

Questions

1. Which inverters are included in the positive feedback loop for this circuit?

2. Which component in this circuit has the greatest amount of influence on the operating frequency?

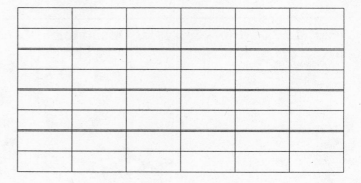

Figure 29-4

Critical Thinking for Project 29

1. One of the advantages of crystal oscillators is that their operating frequency can be changed by simply changing the crystal. Explain what has to be done to an *LC* oscillator, such as a Colpitts or Hartley oscillator, to change its operating frequency.

2. Explain why it is important that frequency sources in communications circuits are stable with respect to changes in power supply voltage and output loading.

Project **30**

Tuned Amplifiers

A Prep Project

This project is a computer simulation of a Class-C tuned amplifier. You will:

- Determine the output frequency and voltage gain of a single-stage tuned amplifier.
- Determine the frequency response characteristics of the amplifier.
- Observe the bandpass characteristics of a 2-stage tuned amplifier.

Preparation

Read Frenzel, *Principles of Electronic Communication Systems*, Section 7-3.

Setup Procedure

1. Select Prep Projects from the Project menu.

2. Select Project 30 Tuned Amplifiers from the list of Prep Projects.

Lab Procedure

In both parts of this project, the rf generator is connected to the input of the circuit and to the upper trace of the oscilloscope. This connection provides the input signal as well as a reference signal for the display. The lower trace of the oscilloscope is connected to a simulated probe.

The project uses interactive block diagrams. You can select a test point to be monitored on the lower trace of the scope by clicking the point with the mouse. The probe is attached to the test point that has the dark blue label.

Part 1   Single-Stage RF Amplifier

Check the project bar across the top of the screen to confirm you are working with Part 1 of this project.

For this part of the project, the rf generator and schematic diagram compete for viewing space on the workbench. Clicking the main body of either one brings it to the foreground for better viewing.

1. Calculate the center frequency (f_c) of the tuned circuit. Record your result on the Results Sheet.

2. Attach the probe to TP 1 of the schematic diagram. Adjust the amplitude of the rf generator to 24 V. Set the rf generator to your calculated center frequency of the collector circuit. Assuming that the vertical sensitivity of the oscilloscope is 24 V/div, record the amplitudes of the rf generator output (Ch A) and the signal at TP 1 (Ch B).

3. Attach the probe to the output of the amplifier at TP 2. Adjust the frequency of the rf generator slightly to make sure you are at or very close to the center frequency.

4. Working from the peak voltage level at TP 2, calculate the upper and lower cut-off voltage levels. Record these levels on the Results Sheet.

5. While monitoring the output at TP 2, adjust the input frequency to the cut-off voltage levels and record the corresponding frequencies on the Results Sheet.

Click the browse button to go to Part 2 of this project.

Part 2   Two-Stage RF Amplifier

Check the project bar to confirm you are working with Part 2 of this project.

1. Fix the probe to TP 1 on the interactive diagram. Adjust the amplitude of the rf generator to 24.0 V.

2. Move the probe to TP 2, and adjust the frequency of the rf generator to obtain a peak voltage level from tuned amplifier 1. Record the frequency from the rf generator and the voltage appearing at TP 2. (Assume the oscilloscope is scaled at 24 V/div.)

3. Determine, by measurement, the upper and lower cut-off frequencies of tuned amplifier 1. Record your findings on the Results Sheet.

4. Move the probe to the output of tuned amplifier 2 (TP 3). Determine the center frequency, and upper and lower cut-off frequencies, for the output at TP 3. Record your data on the Results Sheet.

Click the exit button when you are ready to leave this project.

Results Sheet

Project 30

Part 1   Single-Stage RF Amplifier

Step 1

$f_c =$ _____

Step 2

rf generator voltage = _____

TP 1 voltage = _____

Step 3

rf generator voltage = _____

TP 2 voltage = _____

Step 4

Calculated lower cut-off voltage = _____

Calculated upper cut-off voltage = _____

Step 5

Measured lower cut-off frequency = _____

Measured upper cut-off frequency = _____

Questions

1. Why are the two waveforms of Step 1 identical?

2. How do you know in Step 3 that the input is set near or at the center frequency of the amplifier?

3. What is the bandwidth of the circuit?

Part 2   Two-Stage RF Amplifier

Step 2

Center frequency of
tuned amplifier 1 = _____

Peak voltage of
tuned amplifier 1 = _____

Step 3

Lower cut-off frequency of
tuned amplifier 1 = _____

Upper cut-off frequency of
tuned amplifier 1 = _____

Step 4

Center frequency at TP 3 = _____

Lower cut-off frequency at TP 3 = _____

Upper cut-off frequency at TP 3 = _____

Questions

1. How did you determine the upper and lower cut-off frequencies of tuned amplifier 1 in Step 3?

2. What is the bandwidth of the signal at TP 2? at TP 3?

Critical Thinking for Project 30

1. Describe the similarities between the response curves for an active bandpass filter circuit and a Class-C tuned amplifier.

2. The center frequencies for the amplifiers in a 2-stage tuned amplifier might be slightly different. Why would this discrepancy be intentional?

Project **31**

Tuned Amplifiers

A Hands-On Project

This project uses a single-stage, class-C transistor amplifier that has a tuned circuit as its collector load. You will:

- Construct the circuit.
- Determine the output frequency and voltage gain.
- Determine the frequency response characteristics of the amplifier.
- Tune the amplifier for maximum gain.

Components and Supplies

1	Resistor, 100 Ω
1	Resistor, 100 kΩ
1	Capacitor, 10 nF
2	Capacitor, 100 nF
1	Trimmer capacitor, 20 - 90 pF
1	Inductor, 1 mH
1	NPN transistor, 2N3904 or 2N2222

Equipment

1	Dc power supply
1	Dual-trace oscilloscope
1	Function generator
1	Frequency counter (optional)

Preparation

Read Frenzel, *Principles of Electronic Communication Systems*, Section 7-3.

Complete the work for Prep Project 30.

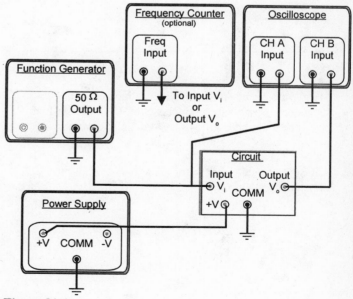

Figure 31-1

Lab Procedure

Part 1   Variable Input Frequency, Fixed Tank Frequency

1. Construct the class-C amplifier circuit shown in Figure 31-2. Connect the function generator, oscilloscope, and optional frequency counter as shown in Figure 31-1. Adjust the function generator for a 500 mV$_{p-p}$ sinusoidal waveform.

2. Calculate the resonant frequency of the *LC* circuit. Record your result on the Results Sheet.

3. For each of the entries in Table 31-1, set the function generator for the given frequency, double-check the value of v_i (500 mV$_{p-p}$), and record the peak-to-peak value of v_o.

4. Within the range of frequencies in Table 31-1, locate the circuit's actual resonant frequency by adjusting the function generator to obtain the peak output signal level at v_o. Record the frequency and output voltage level.

5. Using $v_{o(max)}$ as the 0 dB level, calculate the dB loss for each of the frequencies in Table 31-1. Then plot the circuit's response curve on the two-cycle semilog graph in Figure 31-3.

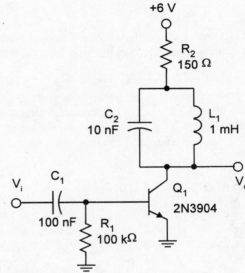

Figure 31-2

Part 2   Fixed Input Frequency, Variable Tank Frequency

1. Modify the circuit in Figure 31-2 by replacing fixed capacitor C_3 with a trimmer capacitor. Connect the function generator and oscilloscope to the circuit as shown in Part 1.

2. Adjust the function generator for a 700 kHz, 500 mV$_{p-p}$ sinusoidal waveform.

3. Carefully adjust the trimmer capacitor, looking for a peak output waveform. Record the value of $v_{o(max)}$ on the Results Sheet.

4. Adjust the amplitude of the function generator at v_i to obtain the largest undistorted signal at v_o. Record the resulting values of v_i and v_o. Calculate and record the voltage gain of the amplifier, based on the voltage values you've just obtained.

Name _____

Results Sheet

Project 31

Part 1   Variable Input Frequency, Fixed Tank Frequency

Step 2

Calculated f_r = _____

Step 4

Measured f_r = _____

Measured $v_{o(max)}$ = _____

f kHz	v_O $V_{p\text{-}p}$	$v_O/v_{i(max)}$	$20 \log(v_O/v_{o(max)})$ dB
6.0			
8.0			
10.0			
30.0			
50.0			
70.0			
90.0			
200.0			
400.0			
500.0			

Table 31-1

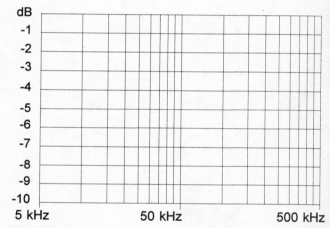

Figure 31-3

Questions

1. What are the values for the upper and lower cut-off frequencies?

2. What is the bandwidth of the circuit?

Part 2   Fixed Input Frequency, Variable Tank Frequency

Step 3

$v_{o(max)} =$ _____

Step 4

$v_i =$ _____ $v_o =$ _____

$A_v =$ _____

Questions

1. What is the voltage gain of this amplifier, expressed in dB?

2. What range of center frequencies are available for this circuit, where the inductor is fixed at 1 mH and C_2 is variable between 20 pF and 90 pF?

Critical Thinking for Project 31

1. Explain how you can tell from the schematic that the amplifier is biased for class-B or class-C operation.

2. Which components in these circuits are responsible for the *flywheel effect*?

3. Explain why this circuit is most efficient at the resonant frequency of the tank circuit.

140

Project **32**

Frequency Multipliers

A Prep Project

This project is a computer simulation of 1- and 2-stage frequency multipliers. You will:

- Note the gain and operating frequencies of a single-stage frequency doubler and tripler.
- Observe the output waveform of a single-stage voltage quadrupler.
- Note the voltage gains, operating frequencies, and waveforms.

Preparation

Read Frenzel, *Principles of Electronic Communication Systems*, Section 7-3.

Setup Procedure

1. Select Prep Projects from the Project menu.

2. Select Project 32 Frequency Multipliers from the list of Prep Projects.

Lab Procedure

This project uses interactive schematic diagrams. Connect the probe to a test point by clicking the test point label.

This simulated probe is connected to a frequency counter and a digital ac meter.

Part 1   Tuned Amplifier as a Frequency Multiplier

Check the project bar across the top of the screen to confirm you are working with Part 1 of this project.

1. Set the amplitude of the rf generator to 10.0 V. Carefully sweep the frequency of the rf generator from its minimum to its maximum limits, watching the readings on the ac digital voltmeter. Note the location of all peaks in this output voltage. For each peak you find (there are five of them), record the frequency setting of the rf generator (f_{in}), the frequency on the frequency counter (f_o), and the voltage on the ac digital voltmeter (v_o).

2. Double-check your results by sweeping the frequency from its maximum to its minimum limit. Resolve any differences you find in the peaks from Step 1.

Click the browse button to go to Part 2 of this project.

Part 2   Single-Stage Frequency Multiplier

Check the project bar across the top of the screen to confirm you are working with Part 2 of this project.

In this part of the project, the input frequency and amplitude are fixed (as they would be when a frequency multiplier is being used for stepping up the frequency of a crystal oscillator circuit).

1. Connect the test probe to TP 2, and repeatedly click the push-button on the filter selector. Note the changes in frequency and amplitude at TP 2.

2. Set the filter selector to the five values shown in Table 32-1. In each case, measure the frequencies and amplitudes at test points TP 1 and TP 2. Also calculate the frequency multiplication factor.

Click the browse button to go to Part 3 of this project.

Part 3   Two-Stage Frequency Multiplier

Check the project bar across the top of the screen to confirm you are working with Part 3 of this project.

Note that the input to this circuit is fixed at 3.5 MHz, 4.8 V_{p-p}.

1. Connect the test probe to the TP 1. Record the output frequency and amplitude on the Results Sheet.

2. Move the test probe to test points TP 2 and TP 3. Record the frequencies and amplitudes that you find.

Click the exit button when you are ready to leave this project.

Results Sheet

Project 32

Part 1   Tuned Amplifier as a Frequency Multiplier

Step 1

First peak $f_i =$ _____ $f_o =$ _____ $v_o =$ _____

Second peak $f_i =$ _____ $f_o =$ _____ $v_o =$ _____

Third peak $f_i =$ _____ $f_o =$ _____ $v_o =$ _____

Fourth peak $f_i =$ _____ $f_o =$ _____ $v_o =$ _____

Fifth peak $f_i =$ _____ $f_o =$ _____ $v_o =$ _____

Questions

1. What is the ratio of $f_i : f_o$ for each peak?

2. What is the dB gain of the circuit for each peak?

Part 2   Single-Stage Frequency Multiplier

Filter Selector	TP 1 f_i	TP 1 v_i	TP 2 f_o	TP 2 v_o	Frequency Multiplier
50 MHz					
100 MHz					
150 MHz					
200 MHz					
250 MHz					

Table 32-1

Questions

1. What is the dB gain of the circuit for each setting?

2. What is the relationship between the amount of frequency multiplication and the voltage gain of this amplifier?

Part 3　　　Two-Stage Frequency Multiplier

Step 1

　　Values at TP 1:

　　　　Frequency = _____　　　Amplitude = _____

Step 2

　　Values at TP 2:

　　　　Frequency = _____　　　Amplitude = _____

　Values at TP 3:

　　　　Frequency = _____　　　Amplitude = _____

Questions

1. What is the frequency multiplication factor for tuned amp 1?

2. What is the voltage gain of tuned amp 1?

3. What is the frequency multiplication factor for tuned amp 2?

4. What is the voltage gain of tuned amp 2?

5. What is the overall frequency multiplication factor of the circuit?

6. What is the overall voltage gain of the circuit?

Critical Thinking for Project 32

1. Explain why the output voltage from a frequency multiplier decreases as the multiplication ratio increases.

2. Resolve the apparent discrepancy between the results you find when working with frequency-multiplier amplifiers and the fact that a pure sinusoidal waveform has no harmonics beyond the fundamental frequency (Project 10).

Project **33**

Frequency Multipliers

A Hands-On Project

This project uses single-stage bipolar transistor amplifier to demonstrate the operation of frequency multipliers. You will:

- Construct the circuit.
- Fine-tune the circuit.
- Apply frequencies and record output frequencies.

Components and Supplies

1	Resistor, 150 Ω
1	Resistor, 100 kΩ
1	Capacitor, 1 μF
1	Capacitor, 100 nF
1	Capacitor, 100 pF
1	Trimmer capacitor, 20 - 90 pF
1	Inductor, 1 mH
1	NPN transistor, 2N3904 or 2N2222

Equipment

1	Variable power supply
1	Function generator
1	Dual-trace oscilloscope
1	Frequency counter (optional)

Preparation

Read Frenzel, *Principles of Electronic Communication Systems*, Section 7-3.

Complete the work for Prep Project 32.

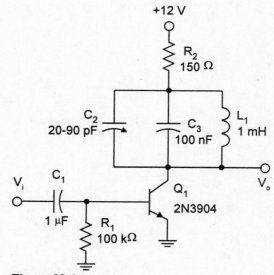

Figure 33-1

Lab Procedure

1. Construct the class-C amplifier circuit shown in Figure 33-1. Connect the function generator to v_i, and adjust it for a 3 V_{p-p} sinusoidal waveform at 455 kHz. Connect the dual-trace oscilloscope to input v_i and to output v_o.

2. Adjust the trimmer capacitor to obtain the peak output signal at v_o. Record the output frequency (f_o) on the first row of Table 33-1.

   Note: The amplifier should now be tuned for maximum gain at 455 kHz.

3. Adjust the frequency of the function generator for 455 kHz/2, or 227.5 kHz, as indicated on the second row of Table 33-1. Record the output frequency (f_o) and ratio of input to output frequency.

4. Repeat the operations in Step Three for the remaining rows in Table 33-1.

   Note: A more stable waveform can be obtained by triggering the horizontal sweep on the oscilloscope from the input signal of the circuit.

5. Sketch the input and output waveforms for the 1:10 frequency-multiplication ratio in Figure 33-2.

Results Sheet

Project 33

Input Frequency (f_i)	Output Frequency (f_o)	Frequency Ratio $(f_i : f_o)$
455 kHz	455 kHz	1 : 1
455 kHz/2		1 : 2
455 kHz/4		
455 kHz/8		
455 kHz/10		

Table 33-1

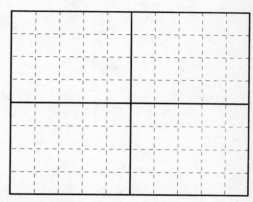

Figure 33-2

Questions

1. Do any of the setups in Table 33-1 represent the operation of a frequency tripler? Explain your answer.

2. What are the minimum and maximum resonant frequencies that can be obtained from the circuit in Figure 33-1, given its values for L_1, C_2, and C_3?

Critical Thinking for Project 33

1. Explain the purpose of the tuned circuit in the collector.

2. Determine the overall multiplication factor of a 4-stage multiplier where the individual stages have frequency multiplication factors of 2, 3, 4, and 6.

3. Explain why a class-C amplifier will always distort a sinusoidal input waveform, and explain why this fact is used to advantage in a frequency multiplier circuit.

Experimental Notes and Calculations

Project **34**

Tuning RF Amplifier Circuits

An Extended Project

This project simulates the behavior of transformer-coupled ac amplifiers. You will:

- Tune the primary and secondary windings of a two-stage, tuned-transformer circuit.
- Tune the windings of a three-stage, tuned-transformer amplifier circuit.

Preparation

Read Frenzel, *Principles of Electronic Communication Systems*, Section 7-3.

Setup Procedure

1. Select Extended Projects from the Project menu.

2. Select Project 34 Tuning RF Amplifier Circuits from the list of Extended Projects.

Lab Procedure

This project uses simulated tunable transformer windings that can be trimmed to produce a peak output at a specified frequency. An rf generator acts as the signal source, and an analog rf voltmeter registers the circuit's output amplitude.

The primary and secondary windings of the transformers are tuned by depressing a mouse key while the mouse pointer is located on a rectangle labeled CW or CCW. This action simulates the turning of the adjustment screw of a

trimmer capacitor or the turning of a threaded powdered-iron slug of a variable inductor. Pressing the left mouse key turns the screw in the counter-clockwise (CCW) direction. Pressing the right mouse key turns the screw in the clockwise (CW) direction. When you "hit the limit" for the screw in one direction, the label turns red and a warning bell sounds. Further adjustment is possible only in the opposite direction.

Part 1   Two-Stage Amplifier

Check the project bar across the top of the screen to confirm you are working with Part 1 of this project.

1. Set the rf generator to 42.5 MHz.

2. Adjust the secondary winding (SEC) to peak the output voltage. Record the approximate indication of the meter.

3. Once again, adjust the primary winding (PRI) to peak the output voltage. Record the approximate indication of the meter.

The coupling transformer is now tuned to 42.5 MHz. In the next series of steps, you will determine the bandwidth of the transformer.

4. Using the voltage reading from Step 3 as the maximum output level, calculate the voltage level for the upper and lower half-power points of the circuit's response curve. Record your values on the Results Sheet.

5. Vary the output of the rf generator to determine the upper and lower cut-off frequencies. Record these frequencies on the Results Sheet.

Click the browse button to go to Part 2 of this project.

Part 2   Three-Stage Amplifier

Check the project bar across the top of the screen to confirm you are working with Part 2 of this project.

1. Set the rf generator for 50.2 MHz.

2. Tune transformer T2 for a peak output voltage. Be sure to tune the secondary winding first. When you have tuned both windings, record the peak output voltage on the Results Sheet.

3. Tune transformer T1 for a peak output voltage (secondary winding, followed by the primary winding). Record the peak output level on the Results Sheet.

4. Carefully tweak the four windings to make sure you are tuned to the highest possible output level. Record this peak output level.

5. Use the voltage of Step 4 as the 0 dB output level, and calculate the voltage output for the upper and lower -3 dB frequencies. Record this voltage level on the Results Sheet.

6. Vary the output frequency of the rf generator to determine the actual upper and lower cut-off frequencies. Record these frequencies on the Results Sheet.

Click the exit button when you are ready to leave this project.

Results Sheet

Project 34

Part 1   Two-Stage Amplifier

Step 2

    Relative voltage out = _____

Step 3

    Relative voltage out = _____

Step 4

    Calculated half-power points voltage = _____

Step 5

    Measured lower cut-off frequency = _____

    Measured upper cut-off frequency = _____

Questions

1. What is the bandwidth of this circuit?

2. What is the Q of this circuit?

Part 2   Three-Stage Amplifier

Step 2

    Relative voltage out = _____

Step 3

    Relative voltage out = _____

Step 4

    Relative voltage out = _____

Step 5

    Calculated half-power points voltage = _____

Step 6

    Measured lower cut-off frequency = _____

    Measured upper cut-off frequency = _____

Questions

1. What is the bandwidth of this circuit?

2. What is the Q of this circuit?

Critical Thinking for Project 34

1. Describe how you would broaden the bandwith of the circuit for Part 1.

2. Prior to the mid-1980s, one common problem with television receivers was a need for routine *IF* alignment. Explain what this problem meant and why it happened.

Experimental Notes and Calculations

Project **35**

Impedance Matching Networks

A Prep Project

This project simulates the operation of an impedance matching network. At a desired frequency, the rf signal source has an internal impedance that is less than the load impedance. You will:

- Show that L-network acts as a parallel tuned circuit when it is matching a lower to a higher impedance.
- Determine the intended operating frequency for a particular L-network.

Preparation

Read Frenzel, *Principles of Electronic Communication Systems*, Section 7-4.

Setup Procedure

1. Select Prep Projects from the Project menu.

2. Select Project 35 Impedance Matching Networks from the list of Prep Projects.

Lab Procedure

This project uses an interactive schematic diagram. The ac voltmeter can be connected to one of the test points, TP 1 or TP 2. You make this selection by clicking the desired test point with the mouse pointer. Also, you can open or close the connection between the signal source and the rest of the circuit. Open or close that connection by clicking the load button, located near the bottom of the schematic diagram. You know the connection is open when you see a break in the conductor between R_g and L.

1. Select TP 1 and open the connection between the signal source and load. This is the no-load condition for the signal source.

2. Set the frequency of the function generator to 200 MHz and the amplitude to each of the settings shown in Table 35-1. For each of these settings, check and record the voltages at TP 1 and TP 2.

3. Set the amplitude of the function generator to 10.0 V and select TP 1. Make sure the load circuit is open. Move the frequency adjustment between its two extremes, and note the effect upon the voltage at TP 1.

4. Click the load button to close the circuit between the signal source and load. Select TP 2 for display on the digital voltmeter, and set the amplitude of the function generator to 10.0 V.

5. Sweep the frequency setting of the function generator until you find a peak reading at TP 2. Record the function generator frequency and voltage reading at TP 2 on the Results Sheet. Select TP 1 and record the voltage found at that point.

6. Make sure the circuit is still closed between the signal source and load, and that the amplitude of the function generator is still set to 10.0 V. Adjust the frequency of the function generator for a frequency that is well away from the center frequency of Step 5. Record this frequency and the voltages at TP 1 and TP 2.

Click the exit button when you are ready to leave this project.

Experimental Notes and Calculations

Results Sheet

Project 35

Step 3

    Note: _____

    _____

Step 5

    Center frequency = _____

    Voltage at TP 2 = _____

    Voltage at TP 1 = _____

Step 6

    Frequency = _____

    Voltage at TP 2 = _____

    Voltage at TP 1 = _____

Amplitude Setting	Voltage TP 1	Voltage TP 2
0		
5		
10		
15		
20		
25		
30		

Table 35-1

Questions

1. How do you account for the results of Step 3?

2. The input and output impedances match at only one frequency. According to the data from this project, what is that frequency?

3. What is the ratio of voltages at TP 1 and TP 2 according to the data of Step 5?

Critical Thinking for Project 35

1. Explain the results shown in Table 35-1.

2. Describe how you would use this project to determine the bandwidth of the impedance-matching effect.

3. Explain why the voltage at TP 1 and TP 2 are equal only at the resonant frequency of the impedance-matching network.

Experimental Notes and Calculations

Project 36

Impedance Matching Networks

A Hands-On Project

The objective of this project is to give you practical experience with *LC* impedance matching networks. You will:

- Construct the circuit.
- Determine the frequency where impedances match.
- Determine changes required for operating the circuit at a different frequency.

Preparation

Read Frenzel, *Principles of Electronic Communication Systems*, Section 7-4.

Complete the work for Prep Project 35.

Components and Supplies

1	Resistor, 300 Ω
1	Inductor, 1 mH
1	Capacitor, 68 nF

Equipment

1	Function generator
1	Dual-trace oscilloscope
1	Frequency counter (optional)

Lab Procedure

The output impedance of a function generator is typically 50 W. The objective of this demonstration is to match this 50 W output with a 300 W load resistance. The type of impedance-matching network used in this project operates to specification only within a range of frequencies. In this example, optimum impedance matching takes place around 17.8 kHz.

1. Construct the external circuit (made up of *L*, *C*, and *R_L*) as shown in Figure 36-1. **Do not** connect this circuit to the function generator at this time, however.

2. Connect the oscilloscope to the output of the function generator and adjust the generator for a sinusoidal waveform at about 17.8 kHz and 10 $V_{p\text{-}p}$. Record the voltage level on the Results Sheet as the unloaded ac output level.

3. Connect the circuit to the function generator to complete the circuit shown in Figure 36-1. Make sure the oscilloscope is still connected across the output of the function generator.

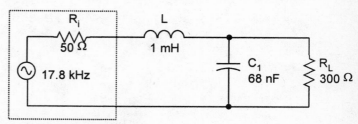

Function Generator

Figure 36-1

4. Carefully adjust the fine tuning of the frequency generator to obtain an output signal that is exactly 1/2 the value of the unloaded output voltage (Step 2). Record this voltage and the frequency at which it occurs.

Experimental Notes and Calculations

Results Sheet

Project **36**

Step 2

Unloaded generator voltage = _____

Step 4

Output voltage at matched impedance = _____

Frequency of matched impedance = _____

Questions

1. There might be a considerable amount of difference between the design frequency for this circuit and the frequency you find in Step 4. How do you account for this discrepancy?

2. What is the phase angle between the current and voltage at R_L when this circuit is properly matched?

Critical Thinking for Project 36

1. Explain the advantage of using a low-pass filter (such as the one used in this circuit) for impedance matching, as opposed to a high-pass filter.

2. Explain why source/load impedance matching can be important for:

    a.   Test instruments
    b.   Transmitters and antennas.

Experimental Notes and Calculations

Project **37**

Simple AM Receiver

A Hands-On Project

This project uses an LM386 audio amplifier as a primitive AM receiver that is similar in principle to the crystal radios used in the early days of radio technology. You will:

- Construct the circuit.
- Make appropriate antenna and ground connections.
- Identify the carrier frequency of the broadcast being heard.

Components and Supplies

1	Capacitor, 100 nF
1	Capacitor, 10 μF
1	IC, LM386 audio power amplifier
1	8 Ω permanent-magnet speaker or earphone
10 ft.	Copper wire, about 18 gauge

The circuit behaves as a detector and audio amplifier, with the detection taking place at PN junctions located at the input terminal of the IC. Since there are no provisions for tuning the circuit, it will respond to the strongest AM broadcast signal in your area.

Preparation

Read Frenzel, *Principles of Electronic Communication Systems*, Section 8-1.

Equipment

1	dc power supply

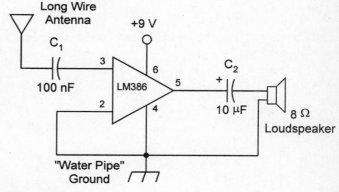

Figure 37-1

Lab Procedure

1. Construct the circuit shown in Figure 37-1.

2. Set up the antenna. Use the long wire as the antenna, stretching it out as long as possible. Make sure you have a good electrical connection at the amplifier end.

3. Make the ground connection. If a literal water-pipe connection is not available, try making a connection to electrical conduit or the metallic faceplate of an electrical outlet. Be sure to connect the COMM of the power supply to this ground as well.

4. Describe your observations of this project on the Results Sheet.

Results Sheet

Project 37

Step 4

_____

_____

_____

_____

_____

_____

Questions

1. What is the carrier frequency for the broadcast you are receiving with this circuit? (Hint: Local newspapers usually publish the broadcast frequency of the AM and FM stations in your area.)

2. What would be the advantage of replacing C_1 with a bandpass filter that could be adjusted for resonant frequencies between 550 kHz and 1650 kHz?

Critical Thinking for Project 37

1. Research and describe the operation of a simple "crystal" radio receiver.

2. Explain why a long antenna and good electrical connection to ground is essential for the operation of this circuit.

Project **38**

Frequency Converter

A Prep Project

The circuit in this project is a frequency converter, or mixer. The input to the circuit is two different frequencies, f_s and f_o. The output is four different frequencies: f_s, f_o, $f_s + f_o$, and $f_s - f_o$. This project gives you a chance to select the amplitude and frequency of waveform f_s and f_o that are applied to the converter circuit. You can then confirm that the output of the circuit consists of the four just cited: the sum, difference, and two original frequencies.

In this project you will:

- Set the frequency and voltage levels for the two inputs of a typical frequency converter circuit.
- Use a tuned circuit and ac voltmeter to determine the frequencies and amplitudes of the signals at the output of a converter circuit.
- Use a simulated spectrum analyzer to confirm the presence of four different frequencies at the output of a converter circuit.

Preparation

Read Frenzel, *Principles of Electronic Communication Systems*, Section 8-3.

Setup Procedure

1. Select Prep Projects from the Project menu.

2. Select Project 38 Frequency Converter from the list of Prep Projects.

Lab Procedure

The procedures you will be using for setting the input amplitudes and frequencies are the same for both Parts 1 and 2 of the project. The difference between the two parts is the type of simulated instruments you will be using for measuring the output frequencies and amplitudes.

In Part 1, the output of the converter circuit is fed to a tuned circuit. The tuned circuit is connected to a digital ac voltmeter. So as you tune the circuit to the four separate output frequencies, you will see readings on the voltmeter rising and falling across peak voltage levels.

In Part 2, you will replace the tuned circuit and ac voltmeter with a spectrum analyzer. In this case, an oscilloscope display clearly shows the four output frequencies and their individual voltage levels.

Part 1   Tuned Output Filter

Check the project bar across the top of the screen to confirm you are working with Part 1 of this project.

1. Set the amplitudes for both rf generators to 12.6 V. Set the output of the rf generator (labeled f_s) to 50.0 MHz. Set the output of the second rf generator (labeled f_o) to 5.0 MHz. Record on the Results Sheet the four frequencies you can expect to find at the output of the mixer circuit.

2. Adjust the tuned filter to select the four frequencies from the mixer circuit. Locate the peaks by noting the maximum voltage levels on the digital ac voltmeter. List the four peak frequencies and their voltage levels in Table 38-1 of the Results Sheet.

Click the browse button to go to Part 2 of this project.

Part 2   Spectral Analysis

Check the project bar to confirm you are working with Part 2 of this project.

Note: The horizontal grid on the spectrum analyzer display is scaled at 10 MHz/div. The vertical grid is scaled at 10 V/div.

1. Set the amplitudes and frequency for both rf generators as specified for Part 1 of this project:

   Rf generator--f_s

         Amplitude: 12.6 V

         Frequency: 50.0 MHz

   Rf generator--f_o

         Amplitude: 12.6 V

         Frequency: 5.0 MHz

2. Sketch the spectrum you see on the analyzer display. Use the grid in Figure 38-1 on the Results Sheet. Label the peaks as f_s, f_o, $f_s - f_o$, and $f_s + f_o$.

3. Determine the frequency and amplitude of each of the peaks on the display. Record your findings in Table 38-2.

4. Set up our own values of frequency and voltage for the inputs to the circuit. Record your values on the Results Sheet.

5. Sketch the spectrum you see on the analyzer display. Use the grid in Figure 38-2 on the Results Sheet.

6. Determine the frequency and amplitude of each of the peaks on the display. Record your findings in Table 38-3.

Click the exit button when you are ready to leave this project.

Name _____

Results Sheet

Project **38**

Part 1   Tuned Output Filter

Step 1

$f_s =$ _____

$f_o =$ _____

$f_s + f_o =$ _____

$f_s - f_o =$ _____

Peak	Frequency	Amplitude
#1		
#2		
#3		
#4		

Table 38-1

Questions

1. Do the results in Step 2 confirm the results anticipated in Step 1?

2. If frequency f_s is increased, what should you do to the frequency f_o in order to maintain the same value for $f_s - f_o$?

Part 2   Spectral Analysis

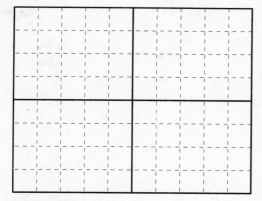

Figure 38-1

Peak	Frequency	Amplitude
#1		
#2		
#3		
#4		

Table 38-2

Step 2

Rf generator--f_s

Amplitude = _____   Frequency = _____

Rf generator--f_o

Amplitude: _____   Frequency: _____

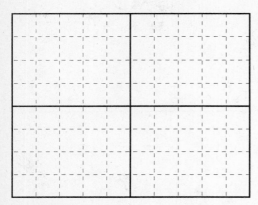

Figure 38-2

Peak	Frequency	Amplitude
#1		
#2		
#3		.
#4		

Table 38-3

Questions

1. Which one of the four peaks is most sensitive to the amplitude of input f_s?

2. Which one of the four peaks is most sensitive to the amplitude of input f_o?

Critical Thinking for Project 38

1. Explain how using a tuned circuit to determine the frequencies at the output of a frequency converter is related to the use of a tuned circuit to select one radio station from all others.

2. Suppose frequency sources f_s and f_o were "ganged" so that the $f_s - f_o$ output always remained at the same frequency. Describe how this would affect the appearance of the four peaks on the spectrum analyzer as you varied one of the frequency sources.

3. Explain why the tuned filter of Part 1 is to be removed before attaching the output of the frequency converter to the spectrum analyzer in Part 2.

Project 39

Frequency Converter

A Hands-On Project

This project uses a germanium diode as the nonlinear element of a frequency converter, or mixer.

- Construct the circuit
- Apply frequencies to be mixed
- Verify the presence of the sum and difference frequencies at the output of the circuit

Preparation

Read Frenzel, *Principles of Electronic Communication Systems*, Section 8-3.

Complete the work for Prep Project 38.

Components and Supplies

2	Resistor, 1 kΩ
2	Resistor, 1.5 kΩ
1	Diode, 1N34 or equivalent
1	455 kHz ceramic filter

Equipment

1	Dual-trace oscilloscope
2	Function generator
1	Frequency counter (optional)

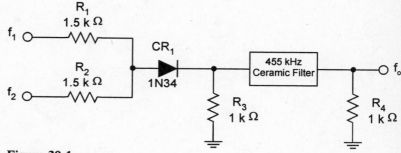

Figure 39-1

Lab Procedure

1. Construct the mixer circuit shown in Figure 39-1. Connect one frequency generator to input f_1 of the circuit, and connect the second to input f_2. Set the output amplitude of both frequency generators to about 6 V_{p-p}. Use sinusoidal waveform outputs for both generators.

2. Using the values for f_1 given in Table 39-1 on the Results Sheet, calculate the values of f_2 such that $f_1 + f_2 = 455$ kHz. Enter your results in the f_2 column.

3. Apply each each combination of f_1 and f_2 in Table 39-1 to the inputs of the circuit. Use the oscilloscope (or frequency counter) to verify the presence of a 455 kHz signal at the output of the circuit.

4. Using the values for f_1 given in Table 39-2, calculate the values of f_2 such that $f_1 - f_2 = 455$ kHz. Enter your results in the f_2 column.

5. Apply each combination of f_1 and f_2 in Table 39-2 to the inputs of the circuit. Use the oscilloscope (or frequency counter) to verify the presence of a 455 kHz signal at the output of the circuit.

6. Set input f_1 to 455 kHz, and vary the frequency of f_2 to 100 kHz above and 100 kHz below f_1. Verify that 455 kH remains as an output frequency.

7. Set input f_2 to 455 kHz, and vary the frequency of f_1 to 100 kHz above and 100 kHz below f_2. Once again, verify that 455 kH remains as an output frequency.

Results Sheet

Project 39

Input f_1 (kHz)	Input f_2 (kHz)	Output f_o (kHz)
100		455
150		455
200		455
250		455
300		455

Table 39-1

Input f_1 (kHz)	Input f_2 (kHz)	Output f_o (kHz)
700		455
750		455
800		455
850		455
900		455

Table 39-2

Step 6

Is 455 kHz present at the output for all values of f_2?

_____

_____

Step 7

Is 455 kHz present at the output for all values of f_1?

_____

_____

Questions

1. What is the function of the portion of the circuit composed of R_1, R_2, and CR_1?

2. What will be the output frequency from this circuit if f_1 and f_2 are both set to 455 kHz?

Critical Thinking for Project 39

1. Cite the similarities in the operation of an AM modulator and a frequency converter.

2. The frequency, 455 kHz, is commonly found in commercial AM radio receivers. What role does it play in these receivers?

Experimental Notes and Calculations

Project **40**

Superheterodyne Receivers

An Extended Project

This project uses simulated electronic instruments and an interactive block diagram to demonstrate the flow of signals through a typical superheterodyne AM radio receiver. In this project you will:

- Trace the flow of signals through the receiver.
- Identify the rf and audio frequencies at critical test points.
- Identify the function of certain parts of the receiver system.

Preparation

Read Frenzel, *Principles of Electronic Communication Systems*, Section 8-5.

Setup Procedure

1. Select Extended Projects from the Project menu.

2. Select Project 40 Superheterodyne Receivers from the list of Extended Projects.

Lab Procedure

Before beginning the actual lab procedure, notice that the project uses two different test instruments: an AM rf generator and a set of frequency counters that measure rf and audio signals.

You can adjust the output of the rf generator by means of the rf frequency slide control. The audio modulating frequency is fixed at 400 Hz, and the percent of modulation is fixed at 100%. Take a moment to experiment with this control.

The measuring instruments located in the upper-right corner of the screen measure rf and audio frequencies at the same time. These two are always connected to the same test point. The rf counter indicates the rf frequency at the test point, and the audio counter indicates the audio frequency at the same point. The only electrical difference between the two counters is that the audio counter has a low-pass filter connected in series with its input.

1. Locate the tuning and band controls on the interactive block diagram. Adjust the tuning control for 1440 kHz.

You have just tuned your simulated AM radio receiver for a station located at 1440 kHz.

2. Likewise adjust the AM rf generator for 1440 kHz.

3. Move the mouse pointer to test point TP 1 on the interactive block diagram. It is found at the antenna.

Notice that the pointer changes from an arrow to a rectangle. This change indicates you have touched TP 1 with the probe for the read-out instruments. Record on the Results Sheet the frequencies you find at TP 1.

4. Move the mouse pointer to TP 2, and record the rf and audio frequencies you find at that point. Then move the pointer to TP 3 and TP 4. Record your figures on the Results Sheet. Be prepared to describe why the readings at these three points differ.

5. Observe and record the frequencies at TP 5 and TP 6. Use the spaces provided on the Results Sheet. Be prepared to describe why the readings at these two points differ.

6. Observe and record the frequencies at TP 7.

7. Make sure the tuning control on the interactive block diagram is still set for 1440 kHz. Adjust the AM rf generator for 1200 kHz. Observe the frequencies at all the test points, and be prepared to explain why they are so different from the results obtained for the first part of this project.

Click the exit button when you are ready to leave this project.

Results Sheet

Project **40**

_____

Step 3

    TP 1 rf frequency = _____

    TP 1 audio frequency = _____

Step 4

    TP 2 rf frequency = _____

    TP 2 audio frequency = _____

    TP 3 rf frequency = _____

    TP 3 audio frequency = _____

    TP 4 rf frequency = _____

    TP 4 audio frequency = _____

According to the block diagram, TP 3 indicates the output of the

_____

According to the block diagram, TP 4 indicates the output of the

_____

Step 5

    TP 5 rf frequency = _____

    TP 5 audio frequency = _____

    TP 6 rf frequency = _____

    TP 6 audio frequency = _____

According to the block diagram, TP 6 indicates the output of the

_____

Step 6

    TP 7 rf frequency = _____

    TP 7 audio frequency = _____

According to the block diagram, TP 7 indicates the output of the

_____

Questions

1. How do you account for the difference in the rf frequency between TP 2 and TP 4, as recorded in Step 4?

2. Why is there such a difference between the readings at TP 5 and TP 6, as recorded in Step 5 of this project?

3. Why is there no rf signal at TP 7, as recorded in Step 6?

4. Which instrument in this circuit is playing the role of a commercial AM broadcast transmitter?

Critical Thinking for Project 40

1. Suppose the tuning of the receiver (interactive block diagram) is set for 1230 kHz. Further suppose the AM rf generator is set for 1430 kHz on the AM band. What will be the signals at TP 1? Describe what you should see at TP 2. Explain these results.

2. Explain how the frequency at TP 3 is always made to be 455 kHz higher than the frequency at TP 1.

3. Explain why there is no audio signal at TP 7 whenever the tuning of the receiver and the frequency of the AM rf generator differ by 5 kHz or more.

Project 41

AM/FM Radio Receiver

An Extended Project

This project uses simulated electronic instruments and an interactive block diagram to demonstrate the flow of signals through a typical AM/FM radio receiver. In this project you will:

- Trace the flow of signals through the receiver when it is operating as an AM receiver.
- Trace the flow of signals when it is operating as an FM receiver.
- Identify the function of certain parts of the receiver system.

Preparation

Read Frenzel, *Principles of Electronic Communication Systems*, Section 8-5.

Setup Procedure

1. Select Extended Projects from the Project menu.

2. Select Project 41 AM/FM Radio Receiver from the list of Extended Projects.

Lab Procedure

Before starting the actual lab procedure, notice that the project uses two different test instruments: an AM/FM rf generator and a frequency counter that measures rf and audio signals at the same time.

You can adjust the output of the rf generator by means of the rf frequency slide control. And you can determine whether the range of frequencies are in the commercial AM or FM broadcast band by pressing the toggle switch on the generator. The audio modulating frequency is fixed at 400 Hz, and the percent of modulation is fixed at 100%. Take a moment to experiment with the slide adjustment and toggle push-button on the rf generator.

The measuring instruments located in the upper-right corner of the screen measure rf and audio frequencies at the same time. These two are always connected to the same test point. The rf counter indicates the rf frequency at the test point, and the audio counter indicates the audio frequency at the same point. The only electrical difference between the two counters is that the audio counter has a low-pass filter connected in series with its input.

There are two kinds of block diagrams visible on the screen at all times. The interactive block diagram clearly shows sections of the AM/FM receiver circuit and the test points for this project. This block diagram is too large to fit on the screen, but you can view all portions by scrolling the diagram to the left or right. Use the horizontal scroll bar near the bottom of the screen for this purpose.

A smaller version of the same block diagram is located near the top of the display. This is the navigation block diagram. Its purpose is to indicate the section of the larger diagram you are viewing at the moment. The blue rectangle on the navigation block diagram indicates where you are on the larger interactive diagram.

Finally, you have access to a small block diagram that shows how the instruments and circuit are interconnected. You can turn this block diagram on and off by clicking the block diagram button located near the upper-right corner of the display.

During this first section of the lab procedure, you will work with the receiver and test instruments in an AM mode.

1. Locate the tuning and band controls on the interactive block diagram. Click the band control for the AM band, and adjust the tuning control for 810 kHz. You have just tuned your simulated AM/FM radio receiver for a station located at 810 kHz on the AM band.

2. Likewise adjust the AM/FM rf generator for 810 kHz of the AM band.

3. Locate test point TP 1 on the interactive block diagram. It is found at the antenna. Scroll the diagram to the left or right if necessary for seeing TP 1. Move the mouse pointer to TP 1. Notice that the pointer changes from an arrow to a rectangle. This change indicates you have touched TP 1 with the probe for the read-out instruments. Note the frequencies. If these output frequencies are not 810 kHz and 400 Hz, check your setup to this point in the project.

4. Move the mouse pointer to TP 2, and record the RF and audio frequencies you find at that point. Use the spaces provided on the Results Sheet. Then move the pointer to TP 3 and record the results. Be prepared to describe why the readings at these two points differ.

5. Observe and record the frequencies at TP 4 and TP 5. Use the spaces provided on the Results Sheet.

6. Observe and record the frequencies at TP 6, TP 7, and TP 8.

7. Observe and record the frequencies at TP 9 and TP 10. Be prepared to describe why the readings at these two points differ.

8. Observe and record the frequencies at TP 11 and TP 12. Be prepared to describe why the readings at these two points are different.

9. Observe and record the frequencies at TP 14 and TP 16.

10. Observe and record the output at TP 17.

During this second phase of the lab procedure, you will work with the receiver and test instruments in the FM mode.

11. Set the tuning and band controls on the interactive block diagram for 90 MHz on the FM band. Then tune the output of the AM/FM rf generator to the same frequency.

12. Observe and record the frequencies at TP 1.

13. Observe and record the frequencies at TP 2 and TP 3. Use the spaces provided on the Results Sheet, and be prepared to explain any differences.

14. Observe and record the frequencies at TP 4 and TP 5.

15. Observe and record the frequencies at TP 6, TP 7, and TP 8.

16. Observe and record the frequencies at TP 9 and TP 10. Be prepared to describe why the readings at these two points differ from those found for the AM mode in Step 7.

17. Observe and record the frequencies at TP 13.

18. Observe and record the output at TP 15.

Click the exit button when you are ready to leave this project.

Results Sheet

Project 41

_____

AM Mode

Step 4

    TP 2 rf frequency = _____

    TP 2 audio frequency = _____

    TP 3 rf frequency = _____

    TP 3 audio frequency = _____

    According to the block diagram, TP 2 indicates the output of the

    _____

    According to the block diagram, TP 3 indicates the output of the

    _____

Step 5

    TP 4 rf frequency = _____

    TP 4 audio frequency = _____

    TP 5 rf frequency = _____

    TP 5 audio frequency = _____

Step 6

    TP 6 rf frequency = _____

    TP 6 audio frequency = _____

    TP 7 rf frequency = _____

    TP 7 audio frequency = _____

    TP 8 rf frequency = _____

    TP 8 audio frequency = _____

Step 7

    TP 9 rf frequency = _____

    TP 9 audio frequency = _____

    TP 10 rf frequency = _____

    TP 10 audio frequency = _____

Step 8

    TP 11 rf frequency = _____

    TP 11 audio frequency = _____

    TP 12 rf frequency = _____

    TP 12 audio frequency = _____

    According to the block diagram, TP 11 indicates an input to a(n)

    _____

Step 9

    TP 14 rf frequency = _____

    TP 14 audio frequency = _____

    TP 16 rf frequency = _____

    TP 16 audio frequency = _____

Step 10

    TP 17 rf frequency = _____

    TP 17 audio frequency = _____

    According to the block diagram, TP 17 indicates the output of the

    _____

FM Mode

Step 12

TP 1 rf frequency = _____

TP 1 audio frequency = _____

Step 13

TP 2 rf frequency = _____

TP 2 audio frequency = _____

TP 3 rf frequency = _____

TP 3 audio frequency = _____

Step 14

TP 4 rf frequency = _____

TP 4 audio frequency = _____

TP 5 rf frequency = _____

TP 5 audio frequency = _____

Step 15

TP 6 rf frequency = _____

TP 6 audio frequency = _____

TP 7 rf frequency = _____

TP 7 audio frequency = _____

TP 8 rf frequency = _____

TP 8 audio frequency = _____

Step 16

TP 9 rf frequency = _____
TP 9 audio frequency = _____
TP 10 rf frequency = _____
TP 10 audio frequency = _____

Step 17

TP 13 rf frequency = _____

TP 13 audio frequency = _____

Step 18

TP 15 rf frequency = _____

TP 15 audio frequency = _____

Questions

1. Why is there such a great difference between the readings at TP 2 and TP 3, as recorded in Step 4 of this project?

2. How do you account for the difference in the rf frequency between TP 4 and TP 5, as recorded in Step 5?

3. Why is there such a difference between the readings at TP 9 and TP 10, as recorded in Step 7 of this project?

4. Why is there no audio signal at TP 11, as recorded in Step 8?

5. Why is there no rf signal at TP 17, as recorded in Step 10?

6. Why is there such a great difference between the readings at TP 2 and TP 3, as recorded in Step 13?

7. Why are the readings for the same two points, TP 9 and TP 10, different for the AM mode (Step 7) and FM mode (Step 16)?

Critical Thinking for Project 41

1. Suppose the tuning on the receiver (interactive block diagram) is set for 1230 kHz on the AM band. Further suppose the AM/FM rf generator is set for 1430 kHz on the AM band. What will be the signals at TP 1? Describe what you should see at TP 2 and TP 3 and explain these results.

2. Suppose the tuning on the receiver and the rf generator are both set for the FM band, but at vastly different frequencies. Describe both the rf and audio signals you should find at TP 1, TP 2, TP 3, and TP 4.

Project **42**

D/A Conversion

A Prep Project

This project simulates the operation of a 4-bit digital-to-analog converter circuit. You will:

- Apply binary values to the input and note the output voltage levels.
- Observe the effect of bias voltage on the output voltage levels.
- Calculate and verify the resolution of the converter.

Preparation

Read Frenzel, *Principles of Electronic Communication Systems*, Section 9-2.

Setup Procedure

1. Select Prep Projects from the Project menu.

2. Select Project 42 D/A Conversion from the list of Prep Projects.

Lab Procedure

The dc voltage source is a variable dc source that has an output range of zero to 23.9 V. In this project, the dc voltage source serves as the reference voltage (V_R) for the DAC.

The digital source provides the digital input for the circuit. This is a simple 4-bit binary source. You can toggle each of the binary values by clicking the corresponding data switch.

The dc voltmeter indicates the actual analog output voltage from the DAC circuit.

1. Adjust the dc voltage source for 15.0 V. Calculate the resolution of this circuit, and enter your response on the Results Sheet.

2. Enter each of the binary values cited in Table 42-1. Record the output voltage reading on the dc voltmeter for each of these values.

3. Graph the data of Table 42-1 in the space provided in Figure 42-1.

   Click the exit button when you are ready to leave this project.

Experimental Notes and Calculations

Results Sheet

Project 42

Step 1

Calculated resolution = _____

D3	D2	D1	D0	v_O
0	0	0	0	
0	0	0	1	
0	0	1	0	
0	0	1	1	
0	1	0	0	
0	1	0	1	
0	1	1	0	
0	1	1	1	
1	0	0	0	
1	0	0	1	
1	0	1	0	
1	0	1	1	
1	1	0	0	
1	1	0	1	
1	1	1	0	
1	1	1	1	

Table 42-1

Questions

1. Based on the data in Table 42-1, what is the actual minimum output voltage?

2. What is the actual maximum output voltage?

3. What is the actual resolution of this circuit?

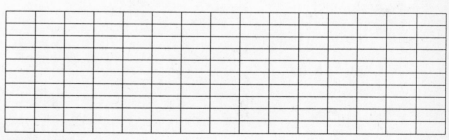

Figure 42-1

Critical Thinking for Project 42

1. Explain why the resolution of an 8-bit D/A converter is inherently greater than the resolution of a 4-bit D/A converter. How much greater is the resolution?

2. Describe what would happen to the resolution of this circuit if the reference voltage were reduced from 15 V to 7.5 V?

Project 43

D/A Conversion

A Hands-On Project

This project demonstrates the operation of a D/A converter circuit using a 4-input R-2R circuit. You will:

- Construct the circuits.
- Determine the analog output voltage level as a function of binary input values.
- Observe and explain the analog output voltage when a high-frequency binary counter is applied to the inputs.

Components and Supplies

5	Resistor, 10 kΩ
4	Resistor, 20 kΩ
1	IC, 741 op-amp
1	IC, 7493 binary counter
4	SPDT switch*

* Jumper wires may be used in place of SPDT switches.

Preparation

Read Frenzel, *Principles of Electronic Communication Systems*, Section 9-2.

Complete the work for Prep Project 42.

Equipment

1	dc power supply
1	Oscilloscope
1	Function generator

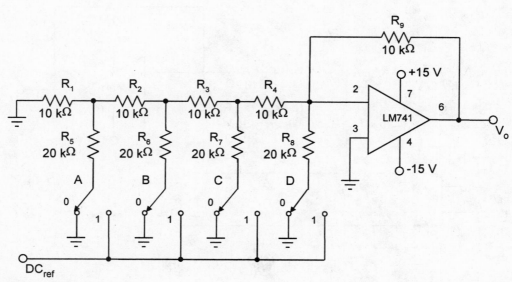

Figure 43-1

Lab Procedure

Part 1   Manual Binary Input

1. Construct the DAC circuit shown in Figure 43-1. Connect the +15 V and -15 V terminals of the power supply to the op-amp IC, and connect the +5 V dc source to the dc_ref line of the circuit.

2. Set all four data switches so that 0 V is applied to all four inputs. Use the oscilloscope to measure the dc level at v_o and record the value in the first line of Table 43-1 on the Results Sheet. Repeat this step for all the logical combinations of binary inputs listed in the table.

3. From your results in Table 43-1, record the minimum and maximum values of output voltage.

   Turn off the power and remove the oscilloscope from the circuit. You will be using the R-2R network and amplifier portion of the circuit in Part 2 of this project.

Part 2   Binary Counter Input

1. Modify the R-2R DAC circuit of Figure 43-1 to replace the manual switches with a 4-bit binary counter (see the complete circuit in Figure 43-2).

2. Adjust the frequency of the function generator to 100 Hz. Adjust oscilloscope to trigger internally on the CLK waveform. Adjust the oscilloscope to show three or four

cycles of the waveform from v_o. Sketch the waveform at v_o in the space provided in Figure 43-3.

3. From the oscilloscope waveform of Step 2, determine the actual minimum and maximum output voltage. Record your findings on the Results Sheet.

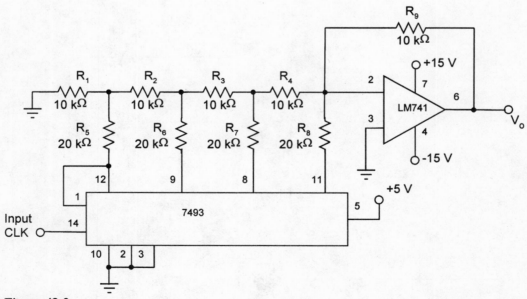

Figure 43-2

Results Sheet

Project 43

Part 1   Manual Binary Input

D3	D2	D1	D0	v_O
0	0	0	0	
0	0	0	1	
0	0	1	0	
0	0	1	1	
0	1	0	0	
0	1	0	1	
0	1	1	0	
0	1	1	1	
1	0	0	0	
1	0	0	1	
1	0	1	0	
1	0	1	1	
1	1	0	0	
1	1	0	1	
1	1	1	0	
1	1	1	1	

Step 3

$v_{O(min)} =$ _____ $v_{O(max)} =$ _____

Questions

1. What is the resolution of the converter in this circuit?

2. Which binary input (D0, D1, D2, or D3) has the greatest amount of influence on the output voltage level?

Table 43-1

Part 2   Binary Counter Input

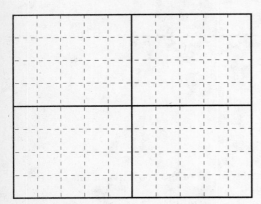

Figure 43-3

Step 3

     Measured $v_{o(min)}$ = _____

     Measured $v_{o(max)}$ = _____

Questions

1. What is the actual resolution of the circuit?

2. How many input clock pulses are required to complete one cycle of the v_o waveform of Figure 43-3?

Critical Thinking for Project 43

1. Explain why it is important that the reference voltage for an R-2R DAC be as stable as possible.

2. A good way to test the operation of a DAC is by applying a binary counting sequence to the data inputs, as done in Part 2 of this project. Cite some circuit troubles and their symptoms that you could detect with this kind of procedure.

Project 44

A/D Conversion

A Prep Project

This project simulates a basic test setup for an 8-bit ADC. You will:

- Apply dc voltage levels to the input of the converter and note the corresponding binary output.
- Calculate the resolution of the circuit, and confirm the result by actual measurement.

Preparation

Read Frenzel, *Principles of Electronic Communication Systems*, Section 9-2.

Setup Procedure

1. Select Prep Projects from the Project menu.

2. Select Project 44 D/A Conversion from the list of Prep Projects.

Lab Procedure

For both parts of this project, the dc voltage source supplies the analog input voltage for the A/D converter circuit. In Part 1, the voltage range is 0 V to 6 V. In Part 2, the voltage range is -12 V to +12 V.

The 8-bit digital display indicates the digital output of the A/D circuit.

Part 1   0 V to 6 V Input

Check the project bar across the top of the screen to confirm you are working with Part 1 of this project.

1. Adjust the dc voltage source for each of the values shown in Table 44-1. In each instance, record the corresponding binary output.

2. Experiment with the circuit to determine how much analog input voltage change is required for changing the binary output by a value of one. Record your finding on the Results Sheet.

3. Adjust the dc voltage source to obtain the binary values cited in Table 44-2. Record the dc voltage source in each case.

Click the browse button to go to Part 2 of this project.

Part 2   -12 V to +12 V Input

The ADC in this part of the project is designed to handle positive and negative analog voltage levels. Such levels are often found in industrial monitoring and control systems.

1. Adjust the dc voltage source for each of the values shown in Table 44-3. In each instance, record the corresponding binary output.

2. Adjust the dc voltage source to obtain the binary values cited in Table 44-4. Record the dc voltage source in each case.

3. Experiment with the circuit to determine its maximum resolution. Record your finding on the Results Sheet.

    Click the exit button when you are ready to leave this project.

Results Sheet

Project **44**

Part 1   **0 V to 6 V Input**

v_i	D7	D6	D5	D4	D3	D2	D1	D0	Decimal Out
0.000									
0.494									
1.012									
1.506									
2.000									
2.494									
3.988									
3.506									
4.000									
4.494									
5.012									
5.506									
6.000									

Table 44-1

Questions

Step 2

Analog output change per input change of
binary 1 = _____

Note: Table 44-2 is located on the next page.

1. What amount of voltage change is required at the input in order to change the digital output by a count of 1?

2. What is the resolution of the circuit measured in terms of units (output) per volt (input)?

D7	D6	D5	D4	D3	D2	D1	D0	Decimal Out	v_i
0	0	0	0	0	0	0	1	1	
0	0	0	0	0	0	1	0	2	
0	0	0	0	0	1	0	0	4	
0	0	0	0	1	0	0	0	8	
0	0	0	1	0	0	0	0	16	
0	0	1	0	0	0	0	0	32	
0	1	0	0	0	0	0	0	64	
1	0	0	0	0	0	0	0	128	

Table 44-2

Part 2   12 V to +12 V Input

v_i	D7	D6	D5	D4	D3	D2	D1	D0	Decimal Out
-12.000									
-10.118									
-8.235									
-6.353									
-4.471									
-2.588									
0.706									
2.118									
4.000									
6.824									
8.706									
10.588									
12.000									

Table 44-3

D7	D6	D5	D4	D3	D2	D1	D0	Decimal Out	v_i
0	0	0	0	0	0	0	1	1	
0	0	0	0	0	0	1	0	2	
0	0	0	0	0	1	0	0	4	
0	0	0	0	1	0	0	0	8	
0	0	0	1	0	0	0	0	16	
0	0	1	0	0	0	0	0	32	
0	1	0	0	0	0	0	0	64	
1	0	0	0	0	0	0	0	128	

Table 44-4

Step 3

Maximum resolution = _____

Questions

1. What amount of voltage change is required for changing the value of the digital output by 1?

2. What is the resolution of the circuit measured in terms of units (output) per volt (input)?

Critical Thinking for Project 44

1. Assume that the circuit used in this project is a successive-approximations converter that is being clocked at 400 kHz. Then calculate the total conversion time.

2. If the circuit used in this chapter happens to be a flash converter, determine the number of comparators it must contain.

Experimental Notes and Calculations

Project 45

A/D Conversion

A Hands-On Project

This project uses 8-bit analog-to-digital IC devices to familiarize you with the A-to-D conversion process. You will:

- Construct the circuit.
- Apply data to the inputs and note the corresponding outputs.
- Calculate output values and confirm the results with actual measurement.

Components and Supplies

1	Resistor, 10 kΩ
1	Potentiometer, 10 kΩ
1	Capacitor, 100 pF
1	IC, 741 op-amp
1	IC, ADC0804 8-bit A/D converter

Equipment

1	Variable dc power supply
1	Function generator
1	Dual-trace oscilloscope
1	Digital voltmeter (optional)

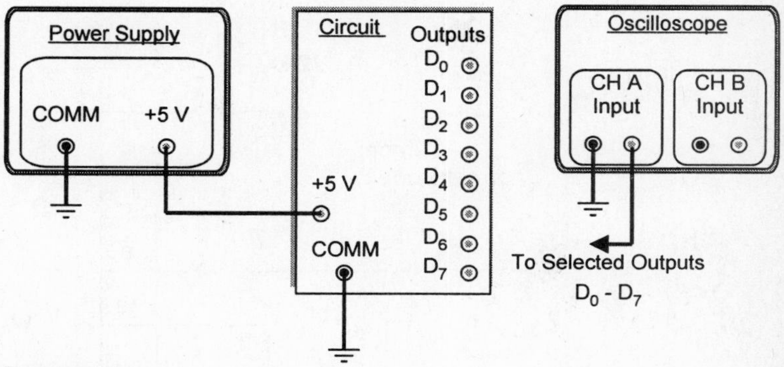

Preparation

Read Frenzel, *Principles of Electronic Communication Systems*, Section 9-2.

Complete the work for Prep Project 44.

Figure 45-1

Lab Procedure

The ADC device specified for this project (ADC0802) is intended for use with a microcomputer bus structure. However, it is used here as a stand-alone device. For this reason, you may find the data outputs latching up occasionally. The operation is easily and reliably restarted by momentarily shorting the connection between pins 3 and 5 to common. This is the purpose of jumper J1 in the circuit.

Special Notes:

- The ADC device is sensitive to static discharge and voltage spikes that might occur while constructing the circuit or changing any part of the circuit while dc power is applied to it. Do not apply power to the ADC circuit in this project until it is fully constructed.
- The dc voltage applied to v_i of the circuit (pin 6 on the ADC) should always be a positive voltage level between zero and +5.1 V dc.

1. Construct the ADC circuit as shown in Figure 45-2. Arrange the laboratory equipment as indicated in Figure 45-1.

2. Adjust the voltage applied to v_i to +5 V and note the dc voltage level at output D_7. If this output is not close to +5 V (logic-1 level), momentarily close the jumper connection between pin 3 and ground.

3. Adjust the voltage applied to v_i between 0 and +5 V, and note the level where output D_7 changes state. Record this level on the Results Sheet.

4. Set the input v_i to the series of voltage levels listed in Table 43-1. In each instance, use the oscilloscope to measure the digital output levels at D_0 through D_7. Record a 0-volt level as 0, and a +5-volt level as 1.

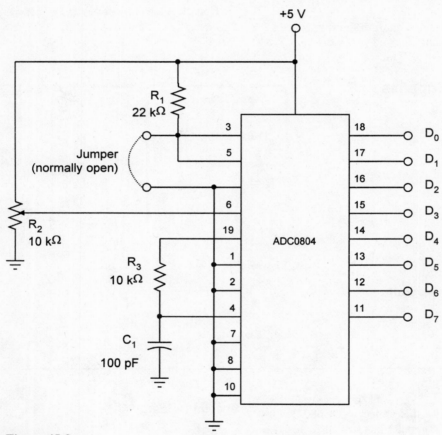

Figure 45-2

Results Sheet

Project 45

Step 3

D_7 changes state when v_i is equal to _____ V.

Questions

1. Does the data in Table 45-1 indicate that the binary value of the output increases with the input voltage level? Explain your answer.

2. What is the theoretical resolution of the ADC in this project?

v_i	D7	D6	D5	D4	D3	D2	D1	D0	Decimal Out
0.000									
0.500									
1.000									
1.500									
2.000									
2.500									
3.000									
3.500									
4.000									
4.500									
5.000									

Table 45-1

Critical Thinking for Project 45

1. Describe a practical way to test the actual resolution of this ADC.

2. Define *quantizing error* and determine the maximum and average error for this circuit.

Project **46**

Pulse-Code Modulation

An Extended Project

This project simulates the operation of an 8-bit pulse-code modulator. You will:

- Apply various dc levels to the input of the modulator and note the sequence of output pulses.
- Determine the output pulse sequence for any amount of dc input.
- Closely observe the operation of a PCM that has a low-frequency sinusoidal waveform applied at its input.

Preparation

Read Frenzel, *Principles of Electronic Communication Systems*, Section 9-4.

Setup Procedure

1. Select Extended Projects from the Project menu.

2. Select Project 46 Pulse-Code Modulation from the list of Extended Projects.

Lab Procedure

The pulse-code modulator in this project simulates the conversion of an analog input to a binary word. The lamps on the modulator indicate the input voltage level. The actual modulation is completed when the binary word is converted to a serial, time-domain signal that appears on the oscilloscope display.

Part 1   Variable dc Input

Check the project bar across the top of the screen to confirm you are working with Part 1 of this project.

1. Adjust the dc voltage source between its two extremes, and note the response on the oscilloscope display. Record the number of pulses appearing on the screen when the input level is minimum and also when it is maximum.

2. Adjust the dc voltage source to the levels indicated in Table 46-1, and record the binary value of the waveform on the oscilloscope.

3. Set the dc voltage source to 18.2 V, and sketch the oscilloscope waveform on the grid in Figure 46-1.

Click the browse button to go to Part 2 of this project.

Part 2   Signal Input

*Check the screen's project bar to confirm you are
working with Part 2 of this project.*

The pulse-code modulator in this part of the project has
a sinusoidal ac voltage applied to its input. While observing the operation of this circuit, answer the questions for Part 2 of the Results Sheet.

*Click the exit button when you are ready to
leave this project.*

Results Sheet

Project 46

Part 1   **Variable dc Input**

Step 1

Number of output pulses with minimum
input level = _____

Number of output pulses with maximum
input level = _____

Questions

1. What is the approximate conversion time of this circuit?

2. What amount of change at the input is causing change at the output?

v_i	D7	D6	D5	D4	D3	D2	D1	D0	Decimal Out
0.0									
2.0									
4.0									
6.0									
8.0									
10.0									
12.0									
14.0									
16.0									
18.0									
20.0									
22.0									
24.0									

Table 46-1

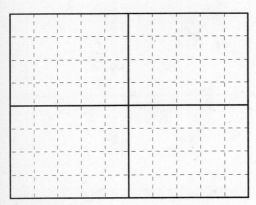

Figure 46-1

Part 2   Signal Input

Questions

1. What is the approximate period of the input ac waveform? the frequency?

2. What is the approximate conversion time of the circuit in this project? (Hint: Note the refresh rate on the oscilloscope display.)

3. What is the lowest binary value at the output of the modulator? the highest?

4. How does the amount of input signal influence the amplitude of the output signal?

Critical Thinking for Project 46

1. Describe the role of analog-to-digital conversion in the process of pulse-code modulation.

2. Cite the portion of this circuit that is displaying serial data and also the portion that is displaying parallel data.

3. Describe how a PCM of this type would be used as part of a telemetry system for continuously monitoring temperatures, fluid levels, rates of gas flow, and other things.

Project **47**

Pulse-Width Modulation

A Prep Project

This project simulates the action of a pulse-width modulator circuit. You will:

- Vary the dc input to a pulse-width modulator and measure the corresponding pulse width and voltage level at the output.
- Plot a performance curve showing output pulse width as a function of the input voltage.
- Determine the correspondence between the instantaneous voltage level of a sinusoidal input and the width of the pulses at the output.

Preparation

Read Frenzel, *Principles of Electronic Communication Systems*, Section 9-5.

Setup Procedure

1. Select Prep Projects from the Project menu.

2. Select Project 47 Pulse-Width Modulation from the list of Prep Projects.

Lab Procedure

Part 1   dc Input

Check the project bar across the top of the screen to confirm you are working with Part 1 of this project.

Notice that the horizontal and vertical scaling for the oscilloscope appears in the upper-right corner of the display. You will need these figures in order to determine the width and amplitude of the circuit's output pulse.

1. Vary the setting of the dc voltage source. Note how varying it affects the width of the pulse on the oscilloscope screen.

2. Set the dc voltage source to the values indicated in Table 47-1. Record the resulting output pulse width and amplitude.

3. Use the results of Table 47-1 to plot a graph showing how the pulse width changes with the amount of input voltage. Use the grid provided in Figure 47-1 on the Results Sheet.

Click the browse button to go to Part 2 of this project.

Part 2   Audio Input

Check the screen's project bar to confirm you are working with Part 2 of this project.

1. Adjust the output of the ac voltage source between its extreme settings. Note the response on the dual-trace oscilloscope display.

2. Set the ac voltage source for a 10 V_{p-p} signal. Sketch the two oscilloscope waveforms on the grid provided in Figure 47-2.

Click the exit button when you are ready to leave this project.

Experimental Notes and Calculations

Results Sheet

Project 47

Part 1   DC Input

Voltage In v_i	Pulse Width Out	Amplitude Out v_o
-8		
-6		
-4		
-2		
0		
2		
4		
6		
8		

Table 47-1

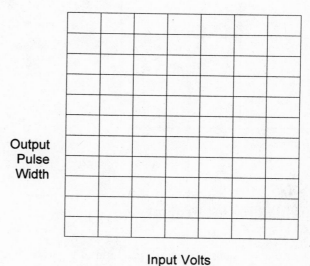

Output Pulse Width

Input Volts

Figure 47-1

Questions

1. According to the graph in Figure 47-1, what is the amount of change in output pulse width per volt of change at the input?

2. According to the graph in Figure 47-1, what output pulse width would you expect upon applying -1 V to the input? +1 V?

3. What is the relationship between the amount of input voltage and the amplitude of the output signal?

Part 2   Audio Input

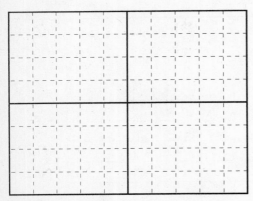

Figure 47-2

Questions

1. As the instantaneous input voltage rises, do the widths of the output pulses increase, decrease, or remain unchanged?

2. As the instantaneous input voltage falls, do the widths of the output pulses increase, decrease, or remain unchanged?

3. As the instantaneous input voltage rises, do the amplitudes of the output pulses increase, decrease, or remain unchanged?

Critical Thinking for Project 47

1. Assuming that the sinusoidal input waveform of Figure 47-2 has a frequency of 100 kHz, determine the approximate sampling frequency of the modulator.

2. Explain how it is possible to use an AM diode detector circuit as a PWM demodulator.

Project 48

Pulse-Width Modulation

A Hands-On Project

This project uses timer circuits to obtain pulse-width modulation. You will:

- Construct the circuit
- Confirm the operation of the two basic parts of the circuit.
- Apply an ac waveform and observe the resulting pulse-width output.

Parts and Materials

2	Resistor, 2.7 kΩ
1	Resistor, 10 kΩ
1	Capacitor, 1μF
1	Capacitor, 10 nF
1	Capacitor, 33 nF
1	Capacitor, 100 nF
2	IC, LM555 timer

Preparation

Read Frenzel, *Principles of Electronic Communication Systems*, Section 9-5.

Complete the work for Prep Project 47.

Equipment

1	dc power supply
1	Function generator
1	Dual-trace oscilloscope

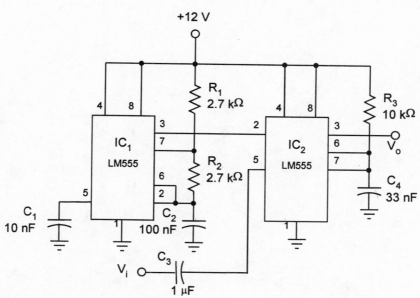

Figure 48-1

Lab Procedure

1. Construct the circuit in Figure 48-1. Connect the oscilloscope to monitor pin 3 of IC_1 on one trace and v_o (pin 3 of IC_2) on the other trace. **Do not** connect the function generator to input v_i at this time.

2. Adjust the sweep rate to 0.1 ms and set the vertical sensitivity of both channels to 5 V/div. Sketch the display in Figure 48-2. Also determine the frequency of both waveforms, and the duration of the positive portion of both waveforms. Record your data on the Results Sheet.

   Note: The waveform from IC_1 is the output of a free-running oscillator. The waveform from IC_2 is the output of a monostable multivibrator that is triggered by the negative-going edge of the oscillator's output pulse.

3. Set the oscilloscope to monitor v_i on one channel and v_o on the other. Trigger the sweep from the signal at v_o. Connect the function generator to input v_i of the circuit, and adjust the generator for a sinusoidal output of about 100 Hz, at 4 $V_{p\text{-}p}$. Leave the sweep rate at 0.1 ms/div.

4. Use the fine-frequency adjustment on the function generator to trim the frequency so that one full cycle of the waveform appears stationary on the oscilloscope display. Note how the duration of the pulses at v_o vary with the instantaneous voltage level at v_i. Make a rough sketch of the display in Figure 48-3. Record the minimum and maximum pulse widths you observe at v_o.

206

Results Sheet

Project 48

Step 2

    Oscillator frequency = _____

    v_o frequency = _____

    Oscillator positive duration = _____

    v_o positive duration = _____

Step 4

    Min pulse width = _____

    Max pulse width = _____

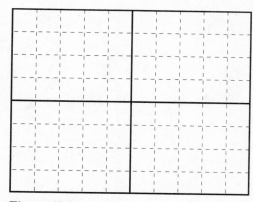

Figure 48-2

Questions

1. According to the data from Step 4 and the display in Figure 48-3, does the output pulse width increase or decrease as the input ac waveform goes more positive?

2. Which IC in this circuit represents the clock section of the PWM? the modulator section? (See Figure 9-30 in the textbook.)

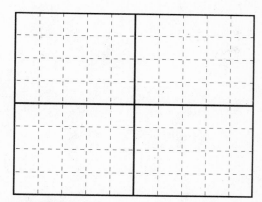

Figure 48-3

Critical Thinking for Project 486

1. The operation of this particular pulse-width modulator depends on a voltage-controlled monostable multivibrator. Explain how this dependence relates to the circuit in this project.

2. Describe the type of circuit that is commonly used for demodulating a PWM waveform.

Experimental Notes and Calculations

Project **49**

Data Multiplexing

A Prep Project

This project uses an interactive block diagram as a tool for studying the flow of data through an 8-channel data multiplexer/demultiplexer system. You will:

- Observe simulated analog data sources at the inputs of an 8-channel multiplexer.
- Relate the binary select code to the input channel that is being transmitted.
- Observe the flow of data through the single data line.
- Relate the binary select code to the output channel that is receiving data.

Preparation

Read Frenzel, *Principles of Electronic Communication Systems*, Section 10-3.

Setup Procedure

1. Select Prep Projects from the Project menu.

2. Select Project 49 Data Multiplexing from the list of Prep Projects.

Lab Procedure

This test setup consists of a 3-bit binary code generator, an interactive block diagram of an 8-channel multiplexer/demultiplexer, and a status board that indicates the nature of the signal at selected test points.

1. Set the binary generator for its manual mode of operation. Set the data switches to the eight different combinations shown in Table 49-1. In each case, click the Load button to transfer the data to the select lines of the multiplexer and demultiplexer.

2. For each of the channel-select codes in Table 49-1, indicate which data input channel is actually being transmitted (Ch 1, Ch 2, Ch 3, and so on). Verify your conclusion by moving the mouse pointer to TP 1 and recording the channel number found at that point.

3. Also for each channel-select code in Table 49-1, indicate which data output channel is actually receiving data.

4. Once again, set the data switches on the binary code generator to the eight combinations of codes shown in Table 49-2. In this case, however, move the mouse pointer

to each of the data output terminals. Mark channels receiving no data with an X. Mark the channel that is receiving data with the channel number being received (Ch 1, Ch 2, Ch 3, and so on) .

5. Click the mode button on the binary code generator for the count-up mode of operation. Click the rate button for the slow counting speed. Use the mouse pointer to observe each data input terminal, and answer Question 1 on the Results Sheet.

6. Use the mouse pointer to observe the data at TP 1 for several complete transmission cycles. Describe what you observe in Question 2 on the Results Sheet.

7. Use the mouse pointer to observe several data output points for several complete transmission cycles. Describe what you observe in Question 3 on the Results Sheet.

Click the exit button when you are ready to leave this project.

Experimental Notes and Calculations

Results Sheet

Project 49

Select Data			Data Input Channel Being Transmitted	TP 1 Status	Data Output Channel Being Received
D2	D1	D0			
0	0	0			
0	0	1			
0	1	0			
0	1	1			
1	0	0			
1	0	1			
1	1	0			
1	1	1			

Table 49-1

Data Output Status

D2	D1	D0	Ch 0	Ch 1	Ch 2	Ch 3	Ch 4	Ch 5	Ch 6	Ch 7
0	0	0								
0	0	1								
0	1	0								
0	1	1								
1	0	0								
1	0	1								
1	1	0								
1	1	1								

Table 49-2

Questions

1. What kinds of changes, if any, do you see at the data input terminals of this circuit while the binary select counter is running?

2. How would you describe the signals you see at test point TP 1 when the select counter is running?

3. Can the data arriving at any one of the output data terminals be described as steady, as arriving in short bursts, or as missing altogether? Describe your observation in more detail.

Critical Thinking for Project 49

1. This 8-channel multiplexer/demultiplexer requires four interconnecting lines: one data line and three select lines. How many data and select lines are required for a similar 16-channel system? a similar 32-channel system?

2. Describe the advantage of running the select clock at 100 kHz or faster, as opposed to running at a frequency less than 10 Hz (as in this simulation).

3. Describe the effect at the data output if the D_2 select line to the demultiplexer is shorted to a logic-0 level.

Project **50**

Data Multiplexing

A Hands-On Project

This project uses an 8-channel CMOS analog multi-plexer/demultiplexer circuit to demonstrate the operation of time-division multiplexing. You will:

- Construct the circuit.
- Apply various waveforms to the inputs and note the changes in the multiplexed output.

Preparation

Read Frenzel, *Principles of Electronic Communication Systems*, Section 10-3.

Complete the work for Prep Project 49.

Parts and Materials

2	Resistor, 15 kΩ
4	Resistor, 22 kΩ
1	IC, 4051 8-channel analog multiplexer
2	SPST switch*

* The switches may be replaced with jumper wires.

Equipment

1	Function generator
1	Dual-trace oscilloscope
1	Power supply

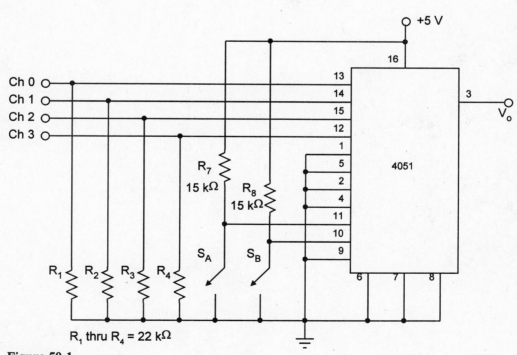

Figure 50-1

Lab Procedure

1. Construct the circuit shown in Figure 50-1. Connect one channel of the oscilloscope to monitor the output of the circuit at v_o.

2. Adjust the function generator for a 100 Hz, 6 V_{p-p}, sinusoidal waveform. Connect the function generator to the Ch 0 input of the circuit. Close switches S_A and S_B. This should send the signal at Ch 0 to the output of the multiplexer. Note the output of the circuit at v_o.

3. Connect the function generator to the Ch 3 input, and open both of the select switches. This action should send the signal at Ch 3 to the output of the multiplexer. Note the output of the circuit at v_o.

4. Set switches S_A and S_B to the four combinations of positions shown in Table 50-1. (For this circuit, closing a switch sets the logic-0 level, and opening the switch creates the logic-1 level.) In each case, connect the signal generator to each of the channels.

Results Sheet

Project 50

Select Data		Data Input Channel at v_o
S_B	S_A	
0	0	
0	1	
1	0	
1	1	

Table 50-1

Questions

1. Is the information being multiplexed in this project considered analog or digital data?

2. Is the output of this circuit considered analog or digital data?

Critical Thinking for Project 50

1. How many inputs are available to this circuit if S_B is shorted to ground? Which inputs would be available?

2. How many select inputs would be required for an 8-channel TDM? 16-channel?

Experimental Notes and Calculations

Project **51**

Parity Generator/Checker

A Prep Project

This project is a computer simulation of odd/even parity generators and checkers. You will:

- Observe and test the operation of a 4-bit parity generator.
- Note the difference between odd and even parity.
- Observe and test the operation of a parity checker.
- Study the operation of a 4-bit parity generator/checker system.

Preparation

Read Frenzel, *Principles of Electronic Communication Systems*, Section 11-5.

Setup Procedure

1. Select Prep Projects from the Project menu.

2. Select Project 51 Parity Generator/Checker from the list of Prep Projects.

Lab Procedure

Throughout this project, a red lamp indicates a logic-1 level, and a black lamp indicates a logic-0 level. Also, a closed switch indicates a logic-1 level, and an open switch indicates a logic-0 level.

Part 1   Parity Generator

Check the project bar across the top of the screen to confirm you are working with Part 1 of this project.

1. Click the O/E switch to set its CLOSED status.

2. Set the input data switches to generate the 4-bit binary codes shown in Table 51-1. Record the corresponding status of the parity lamp.

3. Click the O/E switch to set its OPEN status.

4. Set the input data switches to generate the 4-bit binary codes shown in Table 51-2. Record the corresponding status of the parity lamp.

Click the browse button to go to Part 2 of this project.

Part 2   Parity Generator/Checker

Make sure you are working with Part 2 of this project, as indicated by the title bar on the screen.

1. Click the O/E switch to set its CLOSED status.

2. Set the input data switches to a series of different settings of your choice. Note that the received data follows your input data exactly. Describe on the Results Sheet the response of the parity alarm lamp for all of your data settings.

3. Click the O/E switch to set its OPEN status.

4. Again, set the input data switches to a series of different settings of your choice. Note that the received data follows your input data exactly. Describe on the Results Sheet the response of the parity alarm lamp for all of your data settings.

Click the exit button when you are ready to leave this project.

Results Sheet

Project 51

Part 1   **Parity Generator**

D3	D2	D1	D0	Parity
0	0	0	0	
0	0	0	1	
0	0	1	0	
0	0	1	1	
0	1	0	0	
0	1	0	1	
0	1	1	0	
0	1	1	1	
1	0	0	0	
1	0	0	1	
1	0	1	0	
1	0	1	1	
1	1	0	0	
1	1	0	1	
1	1	1	0	
1	1	1	1	

Table 51-1

D3	D2	D1	D0	Parity
0	0	0	0	
0	0	0	1	
0	0	1	0	
0	0	1	1	
0	1	0	0	
0	1	0	1	
0	1	1	0	
0	1	1	1	
1	0	0	0	
1	0	0	1	
1	0	1	0	
1	0	1	1	
1	1	0	0	
1	1	0	1	
1	1	1	0	
1	1	1	1	

Table 51-2

Questions

1. According to data in Table 51-1, is this circuit operating as an odd-parity generator, or an even-parity generator? Explain your answer.

2. According to data in Table 51-2, is this circuit operating as an odd-parity generator, or an even-parity generator? Explain your answer.

Part 2   Parity Generator/Checker

Step 2

The parity error lamp is

_____ ON continuously

_____ OFF continuously

_____ blinking ON and OFF

Step 4

The parity error lamp is

_____ ON continuously

_____ OFF continuously

_____ blinking ON and OFF

Questions

1. What happens to the parity error lamp when the parity generator is operating as an even-parity generator?

2. What happens to the parity error lamp when the parity generator is operating as an odd-parity generator?

Critical Thinking for Project 51

1. Describe why it is important to match an even-parity generator with an even-parity receiver, and an odd-parity generator with an odd-parity receiver.

2. Describe what would be different about the operation of the parity error lamp if the D_0 bit at the receiver was always a 0, regardless of the status of the D_0 bit at the generator.

220

Project **52**

Parity Generator/Checker

A Hands-On Project

This project uses exclusive-OR logic gates to perform odd/even parity operations. The circuit in Part 1 of the project is a 4-bit parity generator that can operate as both an odd- and even-parity generator. This parity generator is then combined with a 4-bit parity checker in Part 2 of the project. You will:

- Construct the circuits.
- Apply binary inputs to a parity generator and observe the corresponding odd- and even-parity outputs.
- Determine the expected parity value of a binary number, and confirm your answer by actual experiment.
- Apply the output of the parity generator to a parity checker and observe correct and faulty data conditions.

Preparation

Read Frenzel, *Principles of Electronic Communication Systems*, Section 11-5.

Complete the work for Prep Project 50.

Parts and Materials

6	Resistor, 150 Ω
1	Resistor, 2.2 kΩ
2	IC, 7486 TTL quad 2-input exclusive-OR gate
1	IC, 7493 TTL binary counter
6	Red LED
1	SPST switch

Equipment

1	dc power supply
1	Function generator

Lab Procedure

Part 1   Parity Generator

1. Construct the circuit shown in Figure 52-1. Connect the TTL output of the function generator to the trigger input of the counter circuit.

2. Set the function generator to produce a 1 Hz waveform. Apply power to the circuit and close the O/E switch.

Carefully note the status of the lamps, and complete the data in Table 52-1. (Reduce the frequency of the function generator if the states are changing too quickly for you to note them accurately.)

3. Open the O/E switch and complete the data in Table 52-2.

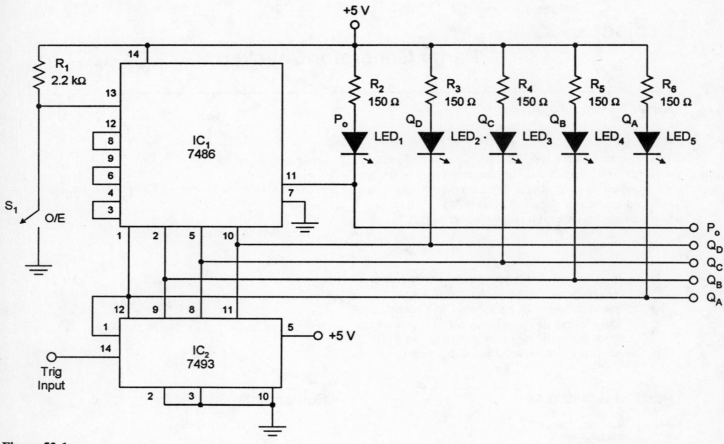

Figure 52-1

Part 2   Parity Checker

1. Construct the 4-bit parity checker circuit shown in Figure 52-2. Attach the data inputs to the outputs of the parity generator as indicated in Figure 52-3. Apply power and the function generator as before.

2. Close the O/E switch in the parity generator portion of the circuit and describe the condition of the parity error lamp as ON continuously, OFF continuously, or blinking ON and OFF.

3. Open the O/E switch in the parity generator portion of the circuit. Describe the condition of the Parity Error lamp as ON continuously, OFF continuously, or blinking ON and OFF.

4. Close the O/E switch. Remove the conductor that connects data line Q_B to IC_3 in the parity checker. (This simulates one type of data corruption that occurs in the transmission of a data signal.) Describe the condition of the Parity Error lamp as ON continuously, OFF continuously, or blinking ON and OFF.

222

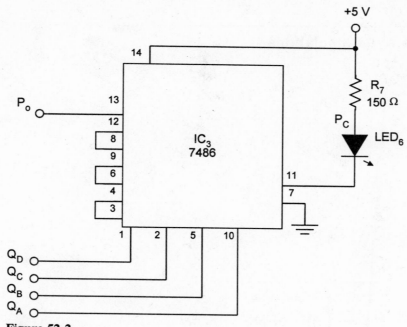

Figure 52-2

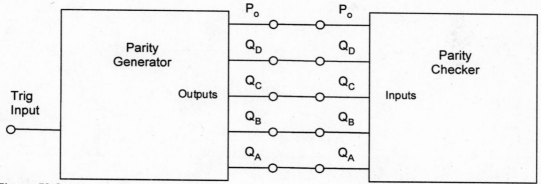

Figure 52-3

Experimental Notes and Calculations

Results Sheet

Project **52**

Part 1   Parity Generator

D3	D2	D1	D0	Parity
0	0	0	1	
0	0	0	0	
0	0	1	1	
0	0	1	0	
0	1	0	1	
0	1	0	0	
0	1	1	1	
0	1	1	0	
1	0	0	1	
1	0	0	0	
1	0	1	1	
1	0	1	0	
1	1	0	1	
1	1	0	0	

Table 52-1

D3	D2	D1	D0	Parity
0	0	0	1	
0	0	0	0	
0	0	1	1	
0	0	1	0	
0	1	0	1	
0	1	0	0	
0	1	1	1	
0	1	1	0	
1	0	0	1	
1	0	0	0	
1	0	1	1	
1	0	1	0	
1	1	0	1	
1	1	0	0	

Table 52-2

Questions

1. According to data in Table 52-1, is the circuit operating as an odd-parity generator or an even-parity generator? Explain your answer.

2. According to data in Table 52-2, is the circuit operating as an odd-parity generator or an even-parity generator? Explain your answer.

Part 2   Parity Checker

Step 2

The condition of the lamp is _____

Step 3

The condition of the lamp is _____

Step 4

The condition of the lamp is _____

Step 5

The condition of the lamp is _____

Questions

1. How do you explain the results of Step 2?

2. How do you explain the results of Step 3?

3. How do you explain the results of Step 4?

Critical Thinking for Project 52

1. Suppose one of the data lines between the parity generator and checker in this project is open, but this fault condition is not visible. Explain how you could use careful observation of the operation of this circuit to determine which line is open.

2. Consider what happens when an even number of data bits are corrupted at the same time. Is it possible that no parity error would occur? Explain your answer.

Project 53

Parallel/Serial Conversion

An Extended Project

This project simulates the action of an 8-bit parallel-to-serial converter. You will:

- Enter 8-bit data into a parallel register.
- Shift the data out of the register serially.

Preparation

Read Frenzel, *Principles of Electronic Communication Systems*, Sections 9-3, 11-2.

Setup Procedure

1. Select Extended Projects from the Project menu.

2. Select Project 53 Serial/Parallel Conversion from the list of Extended Projects.

Lab Procedure

1. Enter binary 11111111 on the parallel input switches, and then press the load button to load the data into the register. Verify that the data loaded, noting the status of the data lamps.

2. Repeatedly press the shift button, and notice how the bits are shifted from left to right.

3. Press the load button to load the same data, then count the number of shift pulses required to send all eight bits through the serial out port. Record the value on the Results Sheet.

4. Load 10000000 into the register and repeatedly click the shift button until the data is completely shifted out.

5. Load 00000001 into the register and repeatedly click the shift button until the data is completely shifted out. Record the number of shift pulses required to send all the data through the serial out port.

Click the exit button when you are ready to leave this project.

Experimental Notes and Calculations

Results Sheet

Project 53

Step 3

Number of shift pulses required = _____

Step 5

Number of shift pulses required = _____

Questions

1. What is the hexadecimal value of the data entered in Step 1? The decimal value?

2. How many serial shift pulses are required to send 00000000?

Critical Thinking for Project 53

1. How many consecutive shift pulses would be required to pass 4-bit data through the serial out port? 16-bit data?

2. If the shift pulses occur at the rate of 320 kHz, how long does it take to send one complete word of 8-bit data?

Experimental Notes and Calculations

Project **54**

Modem

An Extended Project

This project simulates the operation of modems that use quadrature amplitude modulation (QAM). You will:

- Select input data for a QAM and determine the output phase and amplitude of each input.
- Predict from your observations the output phase and amplitude for a given set of binary input data.
- Determine the binary input data required for producing a given QAM phase and amplitude.

Preparation

Read Frenzel, *Principles of Electronic Communication Systems*, Section 11-4.

Setup Procedure

1. Select Extended Projects from the Project menu.

2. Select Project 54 Modem from the list of Extended Projects.

Lab Procedure

The simulated modem device used in this project lets you generate 3- or 4-bit binary values that are immediately translated into a QAM signal. The signal is presented on a special oscilloscope display that directly indicates the signal's amplitude and phase angle.

For experimental purposes, you can select the operating mode for generating the binary data that is fed to the modem. Clicking the mode button cycles the modem through its manual, count-up, count-down, and random modes of op-

eration. When you are using one of the counting modes or the random mode, you can also select the speed at which the binary values change. Do this by clicking the rate pushbutton: slow, medium, and fast.

When you are using the manual mode, click the data buttons to set up the binary value you want to transmit. Then press the load button to see the QAM conversion of that value.

Part 1   8-QAM Modulator

Check the project bar across the top of the screen to confirm you are working with Part 1 of this project.

1. Set up each of the eight binary values listed in Table 54-1. Load each value and record the resulting QAM level on the Results Sheet. Record the phase angle in degrees, and the amplitude as high or low (inner circle or outer circle).

2. Set the modem for the count-up mode and the slow rate. See if you can verify findings you recorded in Table 54-1.

3. Set the modem for the random mode and fast speed. Bear in mind that this is still running far slower than the speeds found in actual QAM modems.

Click the browse button to go to Part 2 of this project.

Part 2   16-QAM Modulator

Check the project bar across the top of the screen to confirm you are working with Part 2 of this project as indicated by the title bar on the screen.

1. Set up each of the sixteen binary values listed in Table 54-2. Load each value and record the resulting QAM level on the Results Sheet. Record the phase angle in degrees, and the amplitude as high or low (inner circle or outer circle).

2. Set up each of the other operating modes for the modem. Observe the responses on the display.

Click the exit button when you are ready to leave this project.

Results Sheet

Project **54**

Part 1   8-QAM Modulator

D2	D1	D0	Phase	Amplitude
0	0	0		
0	0	1		
0	1	0		
0	1	1		
1	0	0		
1	0	1		
1	1	0		
1	1	1		

Table 54-1

Questions

1. Suppose a modem is receiving an 8-QAM signal of 135° high. What binary value does that signal represent?

2. Suppose D_1 of the modem input was always fixed at 0. Which angle-and-level combinations would be missing from the display?

Part 2   16-QAM Modulator

Note: Table 54-2 is located on the next page.

Questions

1. Suppose a modem is receiving a 16-QAM signal of 135° high. What binary value does that signal represent?

2. Suppose a modem is receiving a 16-QAM signal of 247.5° high. What binary value does that signal represent?

D3	D2	D1	D0	Phase	Amplitude
0	0	0	0		
0	0	0	1		
0	0	1	0		
0	0	1	1		
0	1	0	0		
0	1	0	1		
0	1	1	0		
0	1	1	1		
1	0	0	0		
1	0	0	1		
1	0	1	0		
1	0	1	1		
1	1	0	0		
1	1	0	1		
1	1	1	0		
1	1	1	1		

Table 54-2

Critical Thinking for Project 54

1. Explain why the display used in this project is sometimes called a quadrature display.

2. Suppose the modem's low voltage level was higher than it should be. How could this problem be detected on a quadrature display?

3. There is a practical reason for doubling the capacity of a QAM by doubling the number of phase angles rather than the number of different voltage levels. What is that reason?

Project **55**

Antenna Voltage and Current

A Prep Project

This project simulates the behavior of current and voltage for a dipole antenna that is tuned for half-wave resonance. You will:

- Adjust the carrier frequency applied to a half-wave dipole antenna for maximum current.
- Determine the physical length of dipole, based on its half-wave resonant frequency.
- Measure and plot the voltage levels along a dipole antenna that is operating at half-wave resonance.

Preparation

Read Frenzel, *Principles of Electronic Communication Systems*, Section 14-1.

Setup Procedure

1. Select Prep Projects from the Project menu.

2. Select Project 55 Antenna Voltage and Current from the list of Prep Projects.

Lab Procedure

Referring to the block diagram on the screen, note that the rf generator supplies a signal to the dipole antenna.

Part 1   Antenna Current

Check the project bar across the top of the screen to confirm you are working with Part 1 of this project.

In this part of the project, you will use a simulated rf ammeter to monitor the amount of current flowing between the rf generator and the antenna. Note from the block diagram that the meter is connected in series with the transmission line.

1. Set the amplitude of the rf generator to 60 V.

2. Adjust the frequency of the rf generator to determine the half-wave resonant frequency of the antenna. This will be the frequency that causes the largest amount of current to flow to the antenna. Record this frequency on the Results Sheet.

Click the browse button to go to Part 2 of this project.

Part 2   Antenna Voltage

Make sure you are now using Part 2 of this project as noted on the project title bar.

For this part of the project, you will use an rf voltmeter to determine the voltage along the length of a dipole antenna that is operating at half-wave resonance. The amplitude of the signal from the rf generator is adjustable, but the frequency is fixed at 500 MHz for this part of the work.

1. Move the mouse pointer to the antenna figure located on the right-hand side of the screen. Notice how the reading on the voltmeter changes as you move the pointer along the length of the antenna.

2. Set the mouse pointer to the positions indicated in Table 55-1 on the Results Sheet. Be sure you place the pointer directly on the antenna figure, and not on the labels that indicate the positions. Record the voltage levels you find at each of the locations specified in the table.

3. Plot the voltage levels from Table 55-1 in the space provided as Figure 55-1.

Click the exit button when you are ready to leave this project.

Results Sheet

Project **55**

Part 1   Antenna Current

Step 2

Frequency = _____

Questions

1. Based on your data of Step 2, what is the physical length of this half-wave dipole antenna?

2. If the length of this antenna were physically shortened, would you have to increase or decrease the applied frequency in order to restore half-wave resonance?

Part 2   Antenna Voltage

Position	Voltage
0	
$\lambda/16$	
$\lambda/8$	
$3\lambda/16$	
$\lambda/4$	
$5\lambda/16$	
$3\lambda/8$	
$7\lambda/16$	
$\lambda/2$	

Table 55-1

Amplitude
(Relative Volts)

Figure 55-1   Position

Questions

1. Why does this experiment show the same polarity of signal at both ends of the antenna when, in fact, they are of opposite polarity?

2. At what frequency would this antenna be a quarter-wave antenna?

Critical Thinking for Project 55

1. Explain why it is necessary to peak the antenna current (as in Part 1) before you can properly plot its half-wavelength electrical field (as in Part 2).

2. Explain the principle of antenna reciprocity. Suggest how it could apply to the antenna in this project.

Project 56

Antenna Voltage and Current

A Hands-On Project

This project uses an rf signal generator and simple measuring instruments to determine the voltage standing wave ratio of a transmission line and dipole antenna. You will:

- Construct the antenna.
- Tune the system for maximum current.
- Measure relative voltage levels along the length of the antenna.

Preparation

Read Frenzel, *Principles of Electronic Communication Systems*, Section 14-1.

Complete the work for Prep Project 55.

Components and Supplies

1	Capacitor, 100 pF
1	Diode, 1N34 or equivalent
1	dc microammeter, 0 - 50 μA
1	Terminal board, 2-position
2	8-penny nail
6 in.	10 lb nylon fishing line

Wooden board, 1 × 4, 4-1/2 ft long
2 ft. 2 in. 300 Ω twin lead transmission line
4 ft. 6 in. bare copper wire, about 18 gauge

Equipment

1	rf signal generator
1	Tape measure

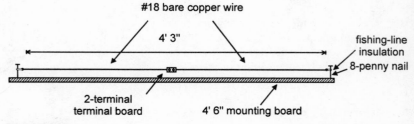

Figure 56-1

Lab Procedure

Part 1   Antenna Assembly

Construct the antenna assembly shown in Figure 56-1. Here are a few hints:

1. Form the two elements of the dipole antenna by cutting the copper wire exactly in half.

2. The critical dimension is the end-to-end length of the copper-wire dipole — it should be very close to the specified length of 4 ft. 3 in.

3. The terminal board should be centered between the ends of the dipole elements.

Part 2   Antenna Current

The real purpose of this part of the project is to make the best possible match between the length of the dipole, the length and characteristic impedance of the transmission line, and the frequency of the rf generator.

1. Connect the transmission line, meter, diode, and rf generator, as shown in Figure 56-2. The meter assembly should be connected in series with the "ungrounded" lead from the rf generator.

2. Set the frequency of the rf generator to about 106 MHz. Adjust the amplitude of the rf signal to a point where the deflection of the meter movement is obvious.

3. Adjust the frequency of the rf generator to peak the antenna current. If the meter pegs before the peak is reached, reduce the signal amplitude. Record the peak frequency on the Results Sheet.

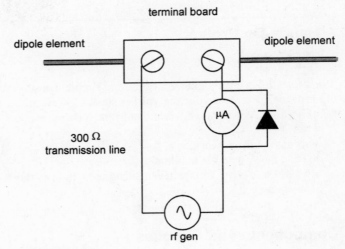

Figure 56-2

Part 3   Antenna Voltage

Important: Part 2 must always be completed prior starting this part. There must be no adjustments to the length of the dipole elements or changes in the frequency setting of the rf generator.

1. Remove the meter assembly from the circuit, and attach both conductors of the transmission line directly to the terminal board and dipole elements. Construct the rf "probe" assembly also shown in Figure 56-3.

2. Complete the ground connection to the "probe" assembly by holding the connection between your fingers. Slide the open end of the capacitor along the elements of the dipole antenna. Note the maximum and minimum responses. Increase the amplitude of the rf generator to get the largest possible amount of response for the maximum readings.

Remember: Do not change the frequency setting from the value established in Part 2 of this project.

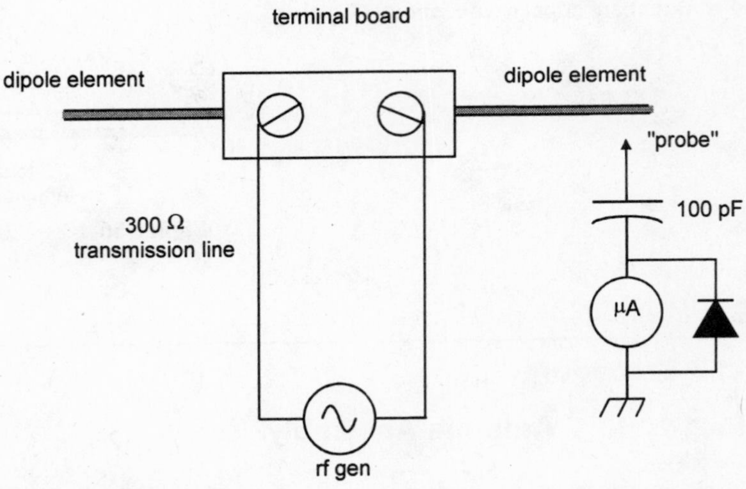

Figure 56-3

3. Locate the "probe" at the outside tip of one of the dipoles. This will be your zero reference location. Record the meter reading in the V column of Table 56-1 on the Results Sheet.

4. Calculate the value of $\lambda/16$ in units of inches. Record your value in the d column of the row labeled $\lambda/16$ on Table 56-1. Locate that point on your antenna, then take a meter reading at that point. Record your results in the V column of the $\lambda/16$ row.

5. Repeat the procedure of Step 4 for all of the locations (except $\lambda/4$) listed in Table 56-1.

6. Plot the results of your measurements on the graph in Figure 56-4.

Experimental Notes and Calculations

Name ______________________________

Results Sheet

Project 56

Part 2 Antenna Current

Step 3

Frequency at peak current = ________

Questions

1. At the frequency of Step 3, is this antenna operating as a full-, half-, or quarter-wave dipole?
2. Given the physical length of this dipole, what is its calculated resonant frequency?
3. How do you account for any difference between the results in Step 3 and your calculation in Question 2?

Part 3 Antenna Voltage

Position	Position *d* in inches	Voltage *V* relative
0		
λ/16		
λ/8		
3λ/16		
λ/4		
5λ/16		
3λ/8		
7λ/16		
λ/2		

Table 56-1

Amplitude (Relative Volts)

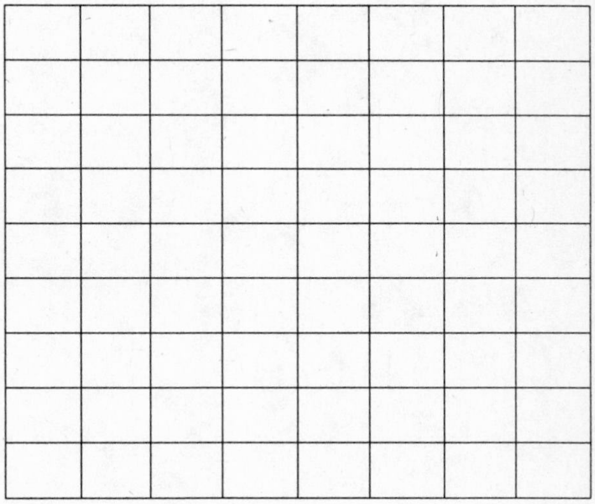

Position

Figure 56-4

Questions

1. What is the purpose of Part 2 of this project?

2. Why is a reading for $\lambda/4$ omitted from Table 56-1?

Critical Thinking for Project 56

1. Explain why it is important to perform Part 2 of this project just prior to performing Part 3.

2. Explain the purpose of the diode that is connected across the terminals of the microammeter.

Project 57

Antenna Impedance Matching

An Extended Project

This project simulates the use of a quarter-wave transformer to match the impedance between a transmission line and antenna. You will:

- Determine the amount of impedance mismatch between a transmission line and antenna.
- Observe the SWR of a transmission line and antenna system.
- Calculate the transformer impedance required for reducing the SWR to 1.
- Adjust the separation of conductors in a quarter-wave transformer to reduce the SWR as close to 1 as possible.

Preparation

Read Frenzel, *Principles of Electronic Communication Systems*, Section 14-2.

Setup Procedure

1. Select Extended Projects from the Project menu.

2. Select Project 57 Antenna Impedance Matching from the list of Extended Projects.

Lab Procedure

This project provides the means for establishing a series of impedance matches and mismatches between a transmission line and antenna. This is done by clicking the buttons located beside the readouts for transmission line and antenna impedance. Digital readouts on the experimental unit also indicate the current values of the quarter-wave transformer and the resulting SWR.

The objective of the work in this project is to vary the impedance of the quarter-wave transformer by adjusting the spacing between the two simulated conductors. This adjustment is made by moving the mouse pointer to one of the elements of the transformer, pressing the left mouse button, and dragging the conductors closer or farther apart. This matching process is completed when you can adjust the transformer to a value of SWR as close to 1 as it can possibly be.

1. Set the line impedance to 50 Ω. Set the antenna impedance to the four values shown in Table 57-1. In each case, adjust the spacing between the conductors of the quarter-wave transformer to obtain the smallest possible SWR. Record the values for this SWR and matching impedance in the table.

2. Set the line impedance to 75 Ω. Repeat the procedure in Step 1, using Table 57-2.

3. Set the line impedance to 150 Ω. Repeat the procedure used in the previous steps. Use Table 57-3 on the Results Sheet.

4. Set the line impedance to 300 Ω. Repeat the procedure used in the previous steps. Use Table 57-4 on the Results Sheet.

Click the exit button when you are ready to leave this project.

Experimental Notes and Calculations

Results Sheet

Project 57

Antenna Impedance	Smallest SWR	Matching Impedance
50 Ω		
75 Ω		
150 Ω		
300 Ω		

Table 57-1

Antenna Impedance	Smallest SWR	Matching Impedance
50 Ω		
75 Ω		
150 Ω		
300 Ω		

Table 57-2

Antenna Impedance	Smallest SWR	Matching Impedance
50 Ω		
75 Ω		
150 Ω		
300 Ω		

Table 57-3

Antenna Impedance	Smallest SWR	Matching Impedance
50 Ω		
75 Ω		
150 Ω		
300 Ω		

Table 57-4

Questions

1. What amount of matching impedance is required when the transmission line impedance already matches the antenna impedance?

2. What kind of impedance match or mismatch is necessary for achieving a value of SWR that is less than 1?

Critical Thinking for Project 57

1. Under certain conditions, the spacing between the conductors of the matching transformer will be less than the spacing of the conductors of the transmission line. Under the opposite set of conditions, the spacing is greater. Describe these conditions and explain why wider or narrower spacing is required.

2. This project deals only with the spacing of conductors in a quarter-wave transformer. Name two other factors that also affect the impedance.

3. Briefly describe why it is important to locate a quarter-wave transformer very close to the antenna.

Project 58

Microwave Systems

A Virtual Project

This project allows you to determine frequency and voltage levels in a simulated microwave transmitter. You will:

- Measure and record voltage and frequencies on an interactive block diagram of a microwave system.
- Determine the frequency multiplication and voltage gain of selected sections of the circuit.

Preparation

Read Frenzel, *Principles of Electronic Communication Systems*, Section 15-1.

Setup Procedure

1. Select Virtual Projects from the Project menu.

2. Select Project 58 Microwave Systems from the list of Virtual Projects.

Lab Procedure

This project uses an interactive block diagram. Clicking a test point on the diagram simulates connecting a test probe to it. This probe is connected to both a frequency counter and an ac voltmeter.

1. Attach the test probe to each of the seven test points as listed in Table 58-1. Record the frequency and voltage amplitude at each point.

2. Answer all the questions included on the Results Sheet.

   Click the exit button when you are ready to leave this project.

Experimental Notes and Calculations

Results Sheet

Project **58**

Test Point	Frequency	Amplitude
TP 1		
TP 2		
TP 3		
TP 4		
TP 5		
TP 6		
TP 7		

Table 58-1

Step 2

1. What is the operating frequency of the oscillator? _____

2. What is the frequency multiplication factor for Amp 1? _____

3. What is the voltage gain, in dB, for Amp 1? _____

4. What is the frequency multiplication factor for Amp 2? _____

5. What is the voltage gain, in dB, for Amp 2? _____

6. What is the frequency multiplication factor for Amp 3? _____

7. What is the voltage gain, in dB, for Amp 3? _____

8. What is the frequency multiplication factor for Amp 4? _____

9. What is the voltage gain, in dB, for Amp 4? _____

10. What is the overall frequency multiplication for Amp 1 through Amp 4? _____

11. What is the overall voltage gain, in dB, for Amp 1 through Amp 4? _____

Critical Thinking for Project 58

1. Explain how you know whether this block diagram represents an AM or an FM transmitter.

2. Cite the primary purpose of amplifiers Amp 1 through Amp 4.

3. Explain how it is possible to see a positive voltage gain through the bandpass filter in this block diagram.

Project **59**

Pulse-Tone Dialer

An Extended Project

This project simulates the operation of a telephone pulse-tone generator. You will:

- Note the two frequencies used for each button on the standard telephone keypad.
- Listen to the frequencies individually.
- Listen to the frequencies as combined for each button on the keypad.

Preparation

Read Frenzel, *Principles of Electronic Communication Systems*, Section 17-1.

Setup Procedure

1. Select Extended Projects from the Project menu.

2. Select project 59 Pulse-Tone Dialer from the list of Extended Projects.

Lab Procedure

Note: This project is most meaningful when using a computer system that is equipped with a sound card capable of playing WAV files.

This workscreen uses two devices: a typical telephone keypad and a form showing the touch-tone matrix. If your computer is equipped with a sound card, you can check out the touch-tone effect by clicking the keys on the keypad. You can hear the individual tones on any DOS/Windows in-

stallation by clicking the frequency labels on the touch-tone matrix form.

Experiment with the various tones and combinations of tones. If a telephone is available nearby, compare the touch tones from the computer with those from an actual telephone.

Click the exit button when you are ready to leave this project.

Results Sheet

Critical Thinking for Project 59

1. Describe why a pair of tones is transmitted for each key, rather than just one tone.

2. Explain why none of the basic frequencies of the touch-tone system is a harmonic of any other of the basic frequencies.

Experimental Notes and Calculations

Project 60

Fiber Optic Communication

A Hands-On Project

This project uses a standard fiber optic demonstration kit. You will:

- Assemble the transmitter and receiver circuits.
- Prepare the fiber optic couplings.
- Connect the transmitter and receiver via the optical fiber.
- Compare the transmitted and received waveforms.

Components and Supplies

1   Fiber optic communication kit, such as Industrial Fiber Optics #IF-E22B.

Specifications for components are included in the instruction manual supplied with the kit.

Preparation

Read Frenzel, *Principles of Electronic Communication Systems*, Sections 18-2, 18-3, and 18-4.

Instruction manual supplied with the fiber optic kit.

Equipment

1	dc power supply
1	Function generator
1	Dual-trace oscilloscope

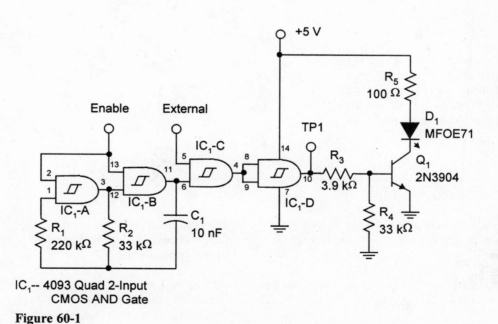

IC₁-- 4093 Quad 2-Input CMOS AND Gate

Figure 60-1

Lab Procedure

Part 1   IR Transmitter and Receiver

1. Assemble the transmitter and receiver circuits shown in Figures 60-1 and 60-2. If you are using the recommended kit, follow the instructions in the manual for mounting components onto the printed circuit boards.

2. Connect the transmitter circuit to the +5 V dc power source and apply power. Connect a jumper from the EN terminal to the negative (or ground) side of the power supply, and the EXT terminal to the positive side. Carefully observe the output of the LED device. You should be able to see a faint red glow, which indicates the LED is turned on. When you switch the EXT jumper to the negative side of the power supply, the LED should turn off.

3. Remove the jumper from the EXT terminal, and replace it with the TTL output of the function generator. Set the frequency of the function generator for a 2 or 3 Hz rectangular waveform, and closely observe the light from the LED. It should be blinking at the rate of the signal from the function generator.

4. Connect the receiver circuit to the +5 V dc source. Connect CHB of the oscilloscope to the DATA output terminal of the receiver. Your setup should resemble the one shown in Figure 60-3. Arrange the two circuit boards so that the LED and phototransistor are facing one another, about a half-inch apart. Adjust the trace on the oscillo-scope to see the waveform at the DATA output terminal. Sketch the waveform in Figure 60-4 on the Results Sheet.

5. Move the transmitter farther away from the receiver, but try to maintain the communication link as indicated by the waveform at the DATA terminal of the receiver. Note how far the transmitter and receiver can be separated before the DATA signal drops off -3 dB or more. Record the approximate distance on the Results Sheet.

6. Move the transmitter to about 1 inch from the receiver and make sure you have a good signal at the DATA terminal. Slide a piece of paper between the LED and phototransistor. Record the result on the Results Sheet.

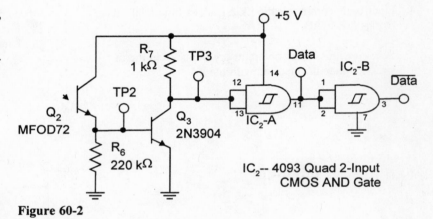

Figure 60-2

Part 2   Fiber-Optic Link

1. Prepare the fiber-optic cable and connectors as described in the instruction manual supplied with the fiber-optic kit.

2. Fasten the cable between the LED on the transmitter circuit and phototransistor on the receiver.

3. With the output of the function generator set for a 2 or 3 Hz rectangular waveform, adjust the oscilloscope to view the receiver waveform at the data terminal on the receiver. Record this waveform in Figure 60-4 on the Results Sheet.

4. Move the transmitter and receiver as far apart as possible, and rotate one of the circuit boards so that the LED and phototransistor are no longer aligned. Describe the effect these changes have upon the quality of the signal at the data terminal.

Results Sheet

Project 60

Part 1   IR Transmitter and Receiver

Step 5

Maximum effective operating distance = _____

Step 6

The effect of blocking the signal is

_____

_____

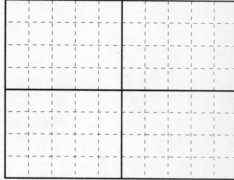

Figure 60-3

Questions

1. Is the modulator in the transmitter circuit an analog or digital type?

2. Is the light-generating device in the transmitter circuit a true laser source?

Part 2   Fiber-Optic Link

Step 4

The effect of moving around the transmitter and receiver is

_____

_____

Questions

1. What is the advantage of using a fiber-optic link as opposed to transmission through air?

2. What type of device is used as the light sensor in the receiver circuit?

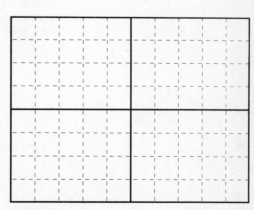

Figure 60-4

Critical Thinking for Project 60

1. Explain the principle of intensity modulation and compare it with rf amplitude modulation.

2. Explain the advantages of using a laser diode as a light source as opposed to a light-emitting diode.

Project 61

Television Remote Controls

A Prep Project

The simulated instruments and circuits in this project include a commercial TV remote control unit, an infrared receiver, and an oscilloscope. In this project your will:

- Observe the sequence of pulses generated by a commercial TV remote control.
- Use an IR receiver and oscilloscope to determine the pulse code pattern of the TV remote control.

Preparation

Read Frenzel, *Principles of Electronic Communication Systems*, Section 19-2.

Setup Procedure

1. Select Prep Projects from the Project menu.

2. Select Project 61 Television Remote Controls from the list of Prep Projects.

Lab Procedure

1. Click the buttons on the TV remote unit at random. Note the type of response appearing on the oscilloscope display. Determine by experiment and observation the number of bits this unit is transmitting. Record your answer on the Results Sheet.

2. Click the buttons in the sequence outlined in Table 61-1. Record the status of each pulse position as *hi* (pulse on) or *lo* (pulse off).

3. Sketch the pulse train for Key 9 in Figure 61-1.

   Click the exit button when you are ready to leave this project.

Experimental Notes and Calculations

Results Sheet

Project **61**

Step 1

Number of bit positions used = _____

Key Pressed	Pulse 1	Pulse 2	Pulse 3	Pulse 4	Pulse 5	Pulse 6	Pulse 7	Pulse 8
0								
1								
2								
3								
4								
5								
6								
7								
8								
9								
VOL UP								
VOL DN								
CH UP								
CH DN								

Table 61-1

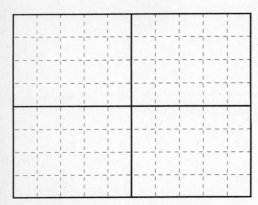

Figure 61-1

Questions

1. If the pulses are 1 ms apart, what is the total period of the longest pulse sequence?

2. If the sweep rate of the oscilloscope is 1 ms/div, what is the width of a hi pulse?

Critical Thinking for Project 61

1. Explain why it is important that the first pulse in each transmission sequence is always a hi pulse.

2. Determine the number of pulse positions required for a TV remote that has 24 different buttons.

3. Earlier TV remote control sometimes used ultrasound rather than IR transmission. What is a major disadvantage of the ultrasound media?

Project **62**

Television Remote Controls

A Hands-On Project

This project provides the opportunity to observe and decode the infrared signal produced by a conventional TV remote control. You will:

- Set up the equipment required for observing the IR signal.
- Determine the number of pulses in the pulse-train sequence.
- Determine the coding that is used for keypad numerals 0 through 9.

Preparation

Read Frenzel, *Principles of Electronic Communication Systems*, Section 19-2.

Complete the work for Prep Project 61.

Components and Supplies

1   IR receiver (from Project 60)
2   TV remote control unit

Equipment

1   Oscilloscope

Lab Procedure

1. Obtain the IR receiver circuit constructed in Project 60. Connect the oscilloscope to monitor the data output of the receiver.

2. Bring the IR emitter window of the TV remote control within an inch of the phototransistor of the IR receiver. Click the buttons on the TV remote at random, adjusting the sweep and amplitude controls on the oscilloscope to show the IR response of the receiver. Set the oscilloscope for triggering. Adjust the trigger slope to show a trace only when one of the buttons on the remote is pressed.

3. Note the type of response appearing on the oscilloscope display. Determine by experiment and observation the number of bits this unit is transmitting. Record your findings on the Results Sheet.

4. Sketch the pulse train for the signal that is transmitted while pressing Key 9. Use the graph provided in Figure 62-1.

5. Press the buttons in the sequence outlined in Table 62-2. Record the status of each pulse position as *hi* (pulse on) or *lo* (pulse off).

Experimental Notes and Calculations

Results Sheet

Project 62

Step 3

Number of pulses per train = _____

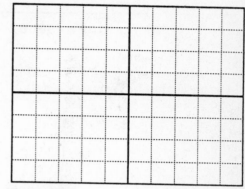

Figure 62-1

Key Pressed	Pulse 1	Pulse 2	Pulse 3	Pulse 4	Pulse 5	Pulse 6	Pulse 7	Pulse 8
0								
1								
2								
3								
4								
5								
6								
7								
8								
9								
VOL UP								
VOL DN								
CH UP								
CH DN								

Table 62-2

Questions

1. What is the time allocated for each pulse? for the entire pulse train?

2. Does your unit automatically repeat the transmission when you hold down a key?

Critical Thinking for Project 62

1. Research and explain the NRZ series code that is commonly used with TV remote controls.

2. Suggest some practical ways to multiply the output power of a TV remote control.

Appendix A

Composite Parts and Equipment Lists

Resistors

All fixed resistors are 1/4 W, 5%

1	Resistor, 100 Ω
1	Resistor, 120 Ω
1	Resistor, 150 Ω
1	Resistor, 180 Ω
1	Resistor, 300 Ω
2	Resistor, 330 Ω
1	Resistor, 470 Ω
1	Resistor, 820 Ω
2	Resistor, 1 kΩ
2	Resistor, 1.5 kΩ
2	Resistor, 2.2 kΩ
1	Resistor, 2.7 kΩ
1	Resistor, 3.3 kΩ
1	Resistor, 4.7 kΩ
5	Resistor, 10 kΩ
4	Resistor, 20 kΩ
1	Resistor, 22 kΩ
1	Resistor, 27 kΩ
1	Resistor, 47 kΩ
1	Resistor, 100 kΩ
1	Resistor, 330 kΩ
1	Potentiometer, 10 kΩ

Capacitors

2	Capacitor, 100 pF
2	Capacitor, 1 nF
2	Capacitor, 10 nF
1	Capacitor, 68 nF
2	Capacitor, 100 nF
1	Capacitor, 1 μF
1	Capacitor, 10 μF
1	Trimmer capacitor, 20 - 90 pF

Inductors

1	Inductor, 1 mH

Discrete Semiconductors

1	Diode, 1N34 or equivalent
1	NPN transistor, 2N3904 or 2N2222
1	JFET, 2N5457

IC Devices

1	IC, LM386 audio power amplifier
2	IC, 555 timer
1	IC, 565 phase-locked loop
1	IC, 566 voltage-controlled oscillator
1	IC, 741 op-amp
1	IC, ADC0804 8-bit A/D converter
1	IC, 3080 operational transconductance amplifier
1	IC, 7404 TTL hex inverter
1	IC, 7493 binary counter
1	IC, 4051 8-channel analog multiplexer

Miscellaneous Electronic Parts

1	455 kHz ceramic filter
1	Crystal, 5.00 MHz
1	Crystal, 3.579 MHz
1	8 Ω permanent-magnet speaker or earphone
1	Fiber optic communication kit, such as Industrial Fiber Optics #IF-E22B

Miscellaneous Supplies

1	Terminal board, 2-position
10 ft.	Copper wire, about 18 gauge
2 ft. 2 in.	300 Ω twin lead transmission line
2	8-penny nail
4 ft. 6 in.	bare copper wire, about 18 gauge
4	SPDT switch
6 in.	10 lb nylon fishing line
	Wooden board, 1 × 4, 4-1/2 ft long

Test Equipment

1	Digital voltmeter (optional)
1	Dual-voltage dc power supply
1	Dual-trace oscilloscope
1	Frequency counter
2	Function generator
1	rf signal generator

Other Equipment

1	AM radio receiver
1	Receiver capable of tuning the 10- or 40-meter shortwave bands
1	TV remote control

Appendix B

Semiconductor Pinouts

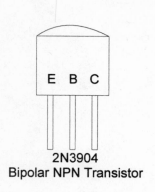

2N3904
Bipolar NPN Transistor

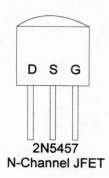

2N5457
N-Channel JFET

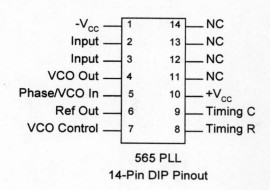

565 PLL
14-Pin DIP Pinout

$-V_{cc}$	1	14	NC
Input	2	13	NC
Input	3	12	NC
VCO Out	4	11	NC
Phase/VCO In	5	10	$+V_{cc}$
Ref Out	6	9	Timing C
VCO Control	7	8	Timing R

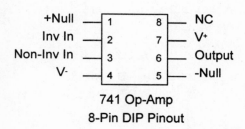

741 Op-Amp
8-Pin DIP Pinout

+Null	1	8	NC
Inv In	2	7	V+
Non-Inv In	3	6	Output
V-	4	5	-Null

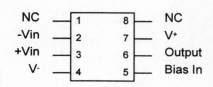

3080 Operational Transconductance Amplifier
8-Pin DIP Pinout

NC	1	8	NC
-Vin	2	7	V+
+Vin	3	6	Output
V-	4	5	Bias In

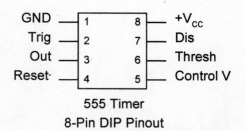

555 Timer
8-Pin DIP Pinout

GND	1	8	$+V_{cc}$
Trig	2	7	Dis
Out	3	6	Thresh
Reset	4	5	Control V

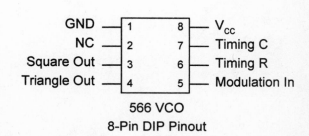

566 VCO
8-Pin DIP Pinout

GND	1	8	V_{cc}
NC	2	7	Timing C
Square Out	3	6	Timing R
Triangle Out	4	5	Modulation In

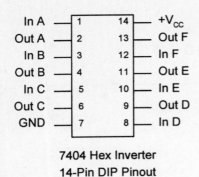

7404 Hex Inverter
14-Pin DIP Pinout

In A	1	14	+V_{CC}
Out A	2	13	Out F
In B	3	12	In F
Out B	4	11	Out E
In C	5	10	In E
Out C	6	9	Out D
GND	7	8	In D

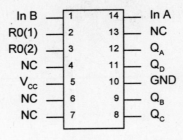

7493 Binary Counter
14-Pin DIP Pinout

In B	1	14	In A
R0(1)	2	13	NC
R0(2)	3	12	Q_A
NC	4	11	Q_D
V_{CC}	5	10	GND
NC	6	9	Q_B
NC	7	8	Q_C

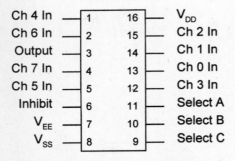

4051 8-Channel Analog Multiplexer
16-Pin DIP Pinout

Ch 4 In	1	16	V_{DD}
Ch 6 In	2	15	Ch 2 In
Output	3	14	Ch 1 In
Ch 7 In	4	13	Ch 0 In
Ch 5 In	5	12	Ch 3 In
Inhibit	6	11	Select A
V_{EE}	7	10	Select B
V_{SS}	8	9	Select C

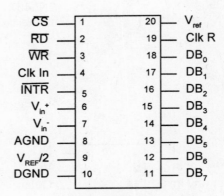

ADC0804 A/D Converter
20-Pin DIP Pinout

$\overline{CS}$	1	20	V_{ref}
$\overline{RD}$	2	19	Clk R
$\overline{WR}$	3	18	DB_0
Clk In	4	17	DB_1
$\overline{INTR}$	5	16	DB_2
$V_{in}+$	6	15	DB_3
$V_{in}-$	7	14	DB_4
AGND	8	13	DB_5
$V_{REF}/2$	9	12	DB_6
DGND	10	11	DB_7